REGINALD JAMES

The Resilient ScaleUp

The Startup CEOs Guide for Scaling Teams Effectively

First Edition: 2023

ISBN Ebook: 978-0-6457577-1-2
ISBN Paperback: 978-0-6457577-2-9

Published by Reginald James
www.reginaldljames.com

Design by Vanessa Mendozzi

Some company and individual's names have been changed to protect the privacy of those involved. The advice and strategies found within may not be suitable for every situation. This work is sold with the understanding that neither the author nor the publisher are held responsible for the results accrued from the advice in this book.

To Mac and Ben. Wherever your paths may lead you, I wish you success and prosperity.

To my wife, Mélanie. My partner for life. Thank you for going "all in" on us. You've taken a tremendous leap of faith coming to Australia. This next season is all for you.

CONTENTS

Introduction

The Foundations of Scale

This isn't another generic book about company culture. In fact, much of what I'm going to tell you spits in the face of famous voices who write about culture. Let's get this straight; I'm not an HR guy, and this is not an HR book. Honestly, I couldn't think of a more boring area of business to talk about. But after founding several companies, working with startups, and helping scaleups build their teams, everything changed for me.

I had just been engaged to transform my client's company, but it was too late. The news broke overnight, and over 400 people were laid off within just a few months. Myst Blue originally began with four founders and a small team of 35. Having been purchased by the large global technology firm ETU Systems and rebranded as ETU Myst Blue in 2008, they began an aggressive scaling journey.

The co-founder, now turned CEO, had only ever led Myst Blue the startup and small business, not Myst Blue the global company with a tremendous growth trajectory. With global customers, the CEO now needed to build out their product and deliver enormous replatforming projects. To do so, he required new talent.

A lot of new talent.

Myst Blue seemingly doubled in size overnight. In the coming years, they would undertake a monumental scaling effort, growing nearly tenfold.

Fast forward ten years and its largest client, Assured Choice, canceled a $450 million project. Having sunk $330 million to date, the project was deemed too costly and was taking too long to continue.

A lot had gone wrong at ETU Myst Blue. Scaling under normal circumstances is difficult. Doing it while navigating an unproven technology can be disastrous if not properly managed. But the company culture created from their recruitment and hiring strategy sealed their fate.

The leaders at ETU Myst Blue miscalculated the waves of change they would experience while scaling. They waited too long to invest in their people and improve their company culture. They falsely believed that their senior managers could make all the tough decisions. They acted as if people would operate as machinery. And they viewed much of their workforce as interchangeable parts. Unsurprisingly, that led the executive team to adopt controversial hiring practices. Using rigorous psychometric testing, Myst Blue eliminated job candidates that were opinionated and most likely to challenge authority. Their preferred employees were those who were compliant to a fault.

Eventually, Myst Blue found itself in a typical product development dilemma. Either build a scalable product for a large market or generate revenues by customizing the solution. Trying to achieve both objectives would lead to significant project delays, cost overruns, and a product that would never scale.

My objective, transforming the company, was therefore doomed to failure. An impossible goal. Myst Blue had scaled their company with the wrong people. Many of those they hired were incapable of adapting.

Scaling your company with the *right* people is one of the significant keys to success. Let's look at another company that puts a high priority on getting the right people.

One Software, a family-owned business, could've never imagined the $25 million exit waiting for them. They helped their financial services customers by selling them loan management software. Star Fall Software, a Canadian multibillion-dollar company, became interested in acquisition, as One Software had grown from $4 million to $10 million in annual revenue.

To support this growth, One Software needed to grow its people. At the time, their team size was just 12. During their scaling journey, they increased that to 36 employees and 35 contractors.

To ensure they brought the right people on board, the leadership team used the Topgrading approach we'll learn about in this book. They developed a three-step interview process that was adopted for all future hiring. Additionally, they measured how well each new candidate matched the requirements of the position using a job scorecard. After conducting competency and culture-based interviews, they consistently brought on the right people.

In *The Resilient Scaleup*, you'll learn about techniques like the Topgrading approach that'll help you hire the right people. You'll get an overview of what it looks like to scale an organization from a team-building perspective.

This book will provide you with the practical insights you need to choose the right recruitment partners and learn which ones to avoid. You'll know why your role as the CEO is critical to the recruitment and team-building process.

While hiring and growing the perfect team is not an exact science, Chapter 7 will explain an often-overlooked method for leveraging luck and serendipity to get the best people on your team. I'll present you with a simple framework to conceptualize the recruitment and growth of your most important people who will build your business for long-term growth and potential exit.

You'll also learn more about your role in cultural development and how it improves talent attraction and management. Most importantly, this book will help you understand how much to invest in your scaleup's culture and when it's the right time to do so.

I've worked with many startups, three of which I co-founded and two of which scaled to employ hundreds of people. While COO of an intelligent assistant scaleup, I found my passion: building and leading high-performing teams.

While growing my consulting business, I helped a property tech scaleup save millions, and possibly their business, by developing a new digital platform that made renting, owning, and managing property stress-free and enjoyable.

In each case, finding and retaining the right people was crucial to our success.

If you're a first-time CEO of a scaleup or someone who doesn't consider themselves a "people person," this book will give you the insights and skills you need as you scale your business.

This book isn't a magic bullet though. It will not solve all the problems you'll face while scaling your business. It's a collection of lessons and business wisdom I've gathered from building teams over the years. At the beginning of each chapter, I'll share some of that wisdom.

The Resilient Scaleup is ultimately about you, the scaling CEO. Building the right team is less about looking outwardly at trends and more about focusing inward and managing the enormous waves of change that come with the job. It isn't about how to create the most incredible company culture that rivals those businesses you know from Palo Alto, California. It's about constantly refining yourself and integrating new lessons and learnings so that you can continually improve and attract better people.

One of the first and most important lessons that all new scaling CEOs must learn is that company culture is overrated. You don't need a company culture like Google or Amazon to attract and retain talent.

› REFERENCES

"How a Software Firm Achieved a $25 Million Exit by Embracing Scaling Up." ScaleUps. February 17, 2022. https://scaleups.com/how-a-software-firm-achieved-a-25-million-exit-by-embracing-scaling-up/.

PART I

THE SCALEUP CULTURE CYCLE

Much has been published about company culture, but not necessarily for businesses in your situation that are scaling up. High-growth companies experiencing rapid expansion are unlike any other type of business. Once your company begins to scale, it will develop a vast array of cultural microcosms peppered throughout the organization; some healthy and some not so healthy. Your challenge is to promote a culture that supports your goals and minimizes the impact of those cultural elements that do not.

Part I of this book consists of a single chapter intended to convince you that a focus on company culture at this stage is still important but often severely overrated. For those who already subscribe to that idea, Chapter 1 will arm you with the behavioral science to further support that belief. It will also outline The Scaleup Culture Cycle, a framework to help you invest in company culture at the right times during your scaling journey.

1

What Scaleups Get Wrong about Culture

Why you don't need a culture like Google or Amazon to attract and retain the best people.

Regardless of your written company values, the organization will be guided by the values of its founders and owners.

On multiple occasions, I consulted with a very large and successful privately owned technology company on leadership and organizational design. This firm called me in because they felt they needed to reorganize themselves. They believed that if they found a better way to structure their departments and organize their projects, they would be much more effective at delivering critical business systems. Their primary problem was that they struggled to deliver their technology projects to the cost and timing expectations of their business stakeholders. This caused their business to hemorrhage cash and angered their investors.

I routinely approached these engagements by conducting a round of interviews. I always began with top leadership and worked my way down to individuals at the coalface. As was often the case, my client's leadership team wanted me to spend most of my time with their people. They wanted me to wave a magic wand and "fix" them. Somehow, I was supposed to miraculously make them more productive.

In interview after interview with their employees, I heard the same stories.

"Our leadership is authoritarian, and I think this company is going in the wrong direction."

"They are so top-down here. This is such a frustrating place to work!"

"They don't listen to us when we tell them about problems with project delivery."

"The leadership style here is command-and-control, and people simply do as they're told."

This wasn't the first time I'd seen this. It seemed to be a reasonably common problem in many of the technology companies I consulted for.

The employees were timid and afraid of their leaders. They could tell me where the problems in the organization had originated, but they were fearful of speaking up. There was a persistent threat of retaliation from their senior managers.

They told me about the different controlling personalities among the executive team. One woman teared up while speaking to me. I could tell these interviews were cathartic for them. Finally, someone was willing to listen to them about the painful realities of working in what felt like a modern-day sweatshop.

Soon, I would share my findings with the CIO. For what it's worth, the CIO was a friendly and personable guy. I found him rather cheerful and easy to talk with. He seemed full of optimism, though I noticed he was frequently missing during our engagement with his team.

What I learned from speaking with him was shocking. Although he was the CIO reporting to the CEO, the power dynamics of that organization didn't follow traditional titles. They rarely do. A powerful group within their new customer experience division had a lot of power and influence, as their business unit brought in the lion's share of the company's overall revenue. They had been unsatisfied with IT for quite some time, and they treated the CIO with the same command-and-control style of leadership that plagued the rest of the company. Because of their lack of faith in him, they kept his department desperately underfunded. Instead, they frequently utilized external, third-party agencies to deliver on their critical initiatives.

Later, I consulted with several more of this company's business divisions. The more I worked with them, the more I heard horrible things about the culture and its leadership. Having done so much work as a transformation consultant, I was pretty used to this sort of thing. But my surprise came a year later when I learned that the company won a prestigious "Best Place to Work in Australia" designation.

I was stunned. Amazed. I barely wanted to consult with that company. It wasn't easy to imagine anyone was truly enjoying working there, especially after all the stories I'd heard.

We've been conditioned to believe that building a successful business requires an outstanding company culture. We're told by leadership experts that we must adopt the latest workplace fads to develop an attractive workplace. However, this was one of several examples where I'd seen the opposite was true. Sure, company culture is important, but at some stage, improving it becomes obsessive.

A Cultural Obsession with Culture

There's no doubt about it. We are a culture-obsessed society. Everywhere you look, people are talking and writing about culture. We can't get enough of it. But what is culture, really? Great Place to Work, a global authority on workplace culture, defines company culture as: "How you do what you do in the workplace." (Lute, 2022) It's the sum of your formal and informal systems, behaviors, and values, all of which create an experience for your customers and employees.

There are hundreds of really great books about company culture that can expand on that definition, so we're not going to discuss that in any more depth here. Besides, most of these books are irrelevant to your scaling organization. That's not to say there isn't useful information there or that you shouldn't bother reading them. However, most authors who write on this theme claim that your company culture is the single most important factor, and unsurprisingly, it's financially advantageous for them to say so.

Recently, the marketplace has exploded with solutions for measuring and improving company culture. One of the latest fads in the last decade is the so-called cultural transformation. Many large enterprise organizations have undertaken massive cultural change programs. We've also seen an explosion of tech businesses promising to improve company culture such as Cultureamp. Cultureamp claims to do this through understanding the employee experience and managing employee performance and development, all through data and analytics.

The foremost thinkers on culture will insist that you constantly improve and build your company culture. They believe this will determine your success in the marketplace. The well-known phrase, "Culture eats strategy for breakfast," (Teasdale, 2002) comes to mind. But perhaps culture isn't everything.

In the 1970s, Firestone, a leading tire brand, had become a victim of its own success. Harvey Firestone Sr. founded the company with the vision that its culture and operations would reflect treating customers and employees as part of the "Firestone family." The Firestone country club was open to all employees, regardless of rank, and Harvey himself maintained close friendships with the top executives of the big carmakers.

Firestone created managers who were fiercely loyal, steeping them in the company's family values and its Akron, Ohio-centered worldview. The company was able to quickly bring new production capacity online by exploiting the booming demand for tires. Frontline employees, for example, would identify market opportunities in the capital-budgeting process and translate them into proposals for investing in additional capacity. Top executives tended to speedily approve the middle managers' recommendations after they selected the most promising proposals.

Firestone's long-standing success gave the company a strong, unified sense of its strategies and values, its relationships with customers and employees, and its operating and investment processes. Since the turn of the century, the company had had a clear formula for success, which served it well.

But, almost overnight, everything changed.

The radial tire was introduced to the U.S. market by Michelin, a French company. Radials were safer, longer lasting, and more economical than traditional bias tires due to a breakthrough in design. When Ford declared in 1972 that all its new cars would have radials, it was clear that they would dominate the U.S. market, effectively decimating the market share of Firestone.

By all accounts, Firestone had built a successful, healthy, family-oriented culture that contributed to its seven decades of success, yet it still failed to maintain its market share. Well, how could this be?

The truth is that familial company cultures often lead to difficulty in making hard decisions that impact employees. While Firestone executives were fully aware of the changing tides in the market, and they did move quickly to invest in a new competitive product, their family-oriented cultural norms impeded them from taking drastic measures. They failed to eliminate underperforming factories, some of which were only operating at 59% capacity. They failed to redesign their production process, which would have meant retrenching large parts of their management and workforce. Firestone clung to their old ways of working.

Survivorship Bias (Being a Copycat)

When they're not obsessing over company culture, these leadership experts are busy promoting the most renowned tech companies as the pinnacle of cultural success. However, as a scaleup business, aiming to become the "best place to work" or develop a company culture like Amazon, Google, or Netflix is doomed to failure. Aspiring to be like any other company with a well-known cultural identity is a poor aspiration. Why? Because it is quite literally the definition of survivorship bias.

One of the earliest examples of factoring survivorship bias into decision-making is from Abraham Wald, a statistician at Columbia University. Wald was born in 1902 in Austria-Hungary. He graduated in mathematics and then lectured in economics in Vienna. After the Anschluss between Nazi Germany and Austria in 1938, Wald and his family faced persecution as Jews and emigrated to the United States after he was offered a university position at Yale, where he joined the Statistical Research Group (SRG).

During World War Two, the U.S. military and the SRG joined forces and attempted to solve problems with a research methodology. The key problem the U.S. military faced was how to reduce the enormous number of aircraft casualties during the war.

The SRG researched the damage their planes received upon returning from conflict. Most of the bullet holes sustained by their planes were in the wings and tail. The engine was not damaged. The U.S. military naturally concluded that the wings and tail were obviously most vulnerable to being hit by bullets.

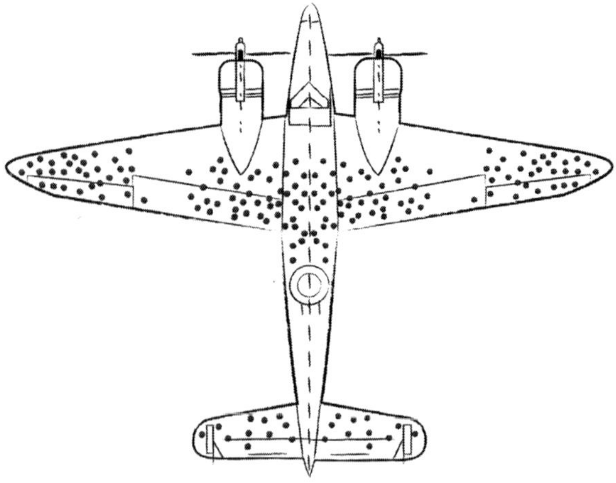

Figure 1.1 An Example of Survivorship Bias

They therefore decided that they needed to increase the amount of armor in these areas. Wald stepped into the room and offered a different perspective. His conclusion was surprising: the wings and tail shouldn't be armored. Shield the engine. Wald's understanding of survivorship bias led him to insightful and reasoned conclusions.

Bias can be defined as any factor in the research process that skews the results. The error of survivorship bias is in looking only at subjects who've reached a certain point without considering the (often invisible) subjects who haven't. The U.S. military were only studying the planes that had returned to base following an encounter with the enemy. Or, simply put, the survivors. What their diagram of bullet holes showed were the areas in which their planes could sustain damage and still be able to fly home, saving their pilots.

Wald surmised that it was actually the engines that were vulnerable. If the engines were hit, the plane and its pilot went down and didn't return to base to be counted in the research. The military listened and armored the engine, not the wings and tail. Over 88,000 casualties were suffered by the U.S. Airforce during the Second World War. If Wald was unaware of survivorship bias, this number would have been higher.

While understandable, our tendency to study and emulate the culture and leadership practices of just a few well-known companies is still driven by survivorship bias. It leads us to an overly simplistic approach to developing culture. We are collectively blind to the multitude of companies that were not successful but have also been great places to work. Our survivorship bias causes us to ignore the companies that prioritized culture, yet no longer exist. Similarly, it causes many culture and leadership experts to ignore otherwise successful firms that don't promote remarkable company culture. By seeking to define your company culture based on the successful methods of a few, you're working without complete data.

In my management consulting days, I was called in to help restructure companies' product and software delivery practices.

During that time, it was popular to be a self-organized, nimble organization. The shining light in this space was Spotify and its way of working, dubbed "the Spotify model." Big, behemoth enterprise firms decided they needed to replicate this approach, filled with its tribes, chapters, guilds, and many other funny-sounding names. Banks like ING and Australia New Zealand Bank hired droves of agile consultants to help them get results like Spotify.

As it turns out, many large banks suffer from deep, systemic issues that prevent them from moving like an already nimble, digital-native company like Spotify. The Spotify model, as confirmed by employees, was no longer a model they even used when people were trying to emulate them. The model only made sense for a season. Spotify's famous tribes, guilds, and chapters were disbanded in favor of a structure that served them better.

Equally interesting was the Netflix employee handbook that became extremely famous on SlideShare, a website that allows its users to share presentations. It stated that there were no maximum number of days an employee could take off. In addition, employees were not required to seek approval for days off. At the time, these practices were uncommon. I've witnessed many companies copy the practices described in Netflix's handbook, which led to detrimental effects. This is because Netflix's mission, its people, its goals, and its collective view of the world are vastly different from any other company. The culture set by Reed Hastings, the CEO, is admirable, but not replicable. Most companies do not have a culture that is psychologically safe enough, supportive enough, or responsible enough to have an unlimited leave policy. Implementing it often leads to worse work-life balance, where employees take less time off than before.

Many scaling businesses fall into this survivorship bias trap. The truth is that you don't need to obsess over your company culture. You're scaling. You're growing. Everything is changing, constantly. The last thing you want is to imitate someone else or be overly focused on the type of company culture you're creating.

The answer isn't to ignore culture, but rather to ensure your focus on company culture gets the appropriate amount of airtime alongside all your other countless responsibilities. To accomplish this, we turn to The Scaleup Culture Cycle.

The Scaleup Culture Cycle

The Scaleup Culture Cycle is a simple framework designed to ensure your business is well-positioned to attract and retain the right people as you go through the different stages of scaling. As I've said before, your business will go through many iterations as you seek to build toward your vision. To ensure you're building the right team, you'll need The Scaleup Culture Cycle.

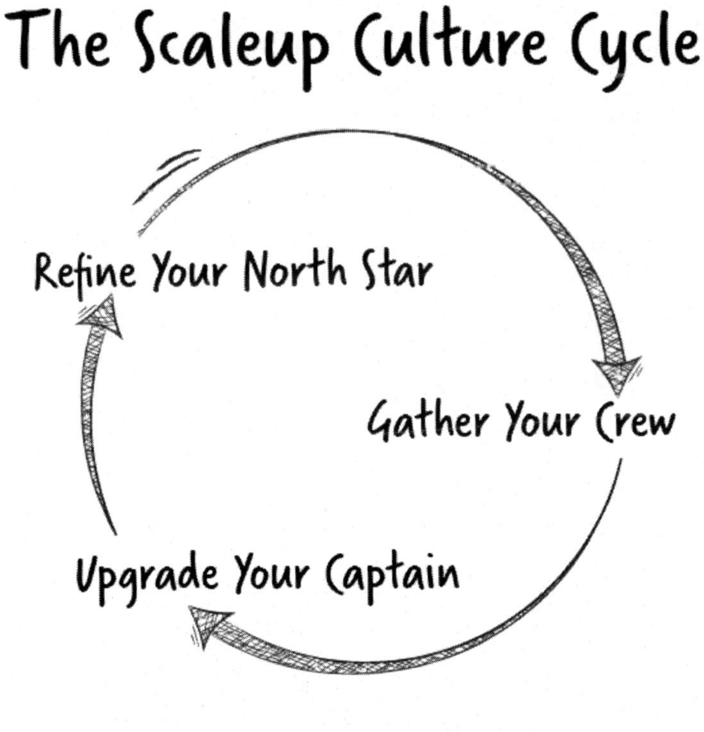

Figure 1.2 The Scaleup Culture Cycle

This book is organized to take you through the three steps of The Scaleup Culture Cycle. I encourage you to reflect on them regularly. Probe deeply to understand where you sit with each one. Throughout your scaling journey, you must continually come back to each component, review them with your immediate leadership team, and systematically plan to ensure your business is ready to transition to the next level. These components are designed to keep you on track.

The first activity is to "refine your North Star." This activity requires you to think very clearly about where you, as a leader, a founder, or CEO of your scaling business, are heading because, whatever the answer, it will form your cultural foundation. It will be your greatest asset or biggest inhibitor to your ability to attract and retain the right people.

The next activity is to "gather your crew." This activity forms the bulk of what you'll focus on in your scaling business. As your business grows and changes, your people are going to have to grow and change with it. And if you cannot change people, you need to *change* the people. You will not be able to take everyone with you into the next phase of your business. As your culture must be refreshed, you must also refresh your workforce. We'll explore this and more related concepts later in the book.

The last activity requires you to "upgrade your captain." This final activity is about improving yourself and your leadership ability.

Some quick takeaways from this chapter:

- Culture is important, but it's not the most important thing at this stage of your business.
- You are different from everyone else. Following the latest fads in leadership and company culture will likely produce many unintended consequences in your scaling business.
- Ignore what everyone else is doing with their company culture and pave your own way.

So many entrepreneurs and business leaders set out to design their companies in the likeness of other great businesses. But being the best is never a matter of comparison. Being the best is not something that popular accolades can define for you.

Being the best begins with knowing where you're going. My guess is that you believe you've already got a clear answer to this. But let's test that clarity. And let's explore how true clarity will empower and strengthen your hiring efforts.

› REFERENCES

Lute, Julian. "Company Culture – Meaning, Benefits and Strategies." Great Place To Work®. Accessed December 16, 2022. https://www.greatplacetowork.com/resources/blog/company-culture-meaning-benefits-and-strategies.

Sull, Donald. "Why Good Companies Go Bad." *Harvard Business Review.* August 1, 2014. https://hbr.org/1999/07/why-good-companies-go-bad.

Teasdale, Sheila. "Culture eats strategy for breakfast!" *Journal of Innovation in Health Informatics* Vol 10, no. 4 (2002):195-196. Accessed January 3, 2023. DOI:10.14236/jhi.v10i4.259

PART II

REFINE YOUR NORTH STAR

One of the most exciting milestones as a business is when you achieve product-market fit. But once you do, an even more exciting journey begins. Expansion. Persistent clarity in this stage of growth is your best friend. Part II of *Scaleup Culture* consists of two chapters, both intended to help you understand your current position as it relates to leadership and culture in your scaling business. These chapters will help you lead intentionally and prepare you to grow your team.

2

Lead the Product

How a powerful vision will energize your hiring efforts and keep your best people committed to you.

Bad leadership is better than no leadership. Without the certainty of direction, people will struggle to be productive.

Without a clear and compelling vision, it's impossible to attract and retain the right people to help you reach your destination, even if the business makes sense to you.

One entrepreneur I worked with had developed a tool for scoring. Or was it a tool for surveying? Or was it a rating system? It's hard to say. This was the classic situation of a solution looking for a problem. That entrepreneur tried to pitch his solution to health professionals in clinical situations to rate the effectiveness of patient care. He tried pitching the solution to education professionals to have children rate their educational experience.

Wave Link was certainly an interesting technical solution because it utilized brand-new gestures functionality in the iPhone's early days. Users could input data in surveys simply by rotating their fingers in a circular motion. But that's really all it was: a novel technology product looking for a problem to solve. Not everyone has the funds to spend on R&D for solutions to unknown problems. In addition to selling Wave Link, a bigger challenge for this entrepreneur was convincing anyone to work on this product.

My team has developed digital products in many organizations, large and small, and we've seen how a powerful vision can elevate a business. Through my resource consultancy, I developed products for a startup with the mission to ensure that every "pickup" basketball team had the right number of team members and that no one was left out. I've developed digital products for the World Wildlife Fund to increase donor engagement and locate a new market of givers. But, even after so many visionary projects, there's one that really stands out.

Scott Bateman was an executive leader in a large property management business based in Melbourne, Australia. Realizing there had been no innovation in the property management space in Australia for over 50 years, he was looking at reinvigorating a property management tool that had led the market at one time but had now fallen out of favor.

In Australia, property management is centered around the owner, which meant costs were driven lower and lower over the years. This ultimately reduced the quality of service, creating a frustrating environment for renters, owners, and property managers. Owners wanted transparency and assurance that their rental properties were being well maintained. Renters faced several challenges, including privacy and scheduling issues when allowing people to come into their property at random times to do inspections.

One of the major stressors came from requiring the renter to be at home to allow maintenance to be completed at the property at odd times. Take, for instance, the all-too-common leaky sink. The tenant had to call the property manager, who then had to call and organize for a plumber to come by. The renter was then required to be home to let the plumber in. Renters frequently had to take time off work for this to happen. Once the plumber discovered the problem, rather than just fixing it, they had to submit a report to the property manager, who subsequently forwarded that report to the owner to seek approval. Often, at this point, the owner got sticker shock and wanted the property manager to get additional quotes to drive the price down. This entire situation created frustration for each party, including the property manager.

Scott's vision for Kolmeo wasn't just to create a new property management system. It was to reinvent the property management profession in Australia by fixing the relationship between the renter, the property owner, and the property manager. He called this relationship "Kolmeo," which is Dutch for "triangle." He saw these relationships as the key to unlocking value and reinventing a dying industry.

We developed the prototype of this new service for Kolmeo. Our team presented it to their board, which unlocked about $14 million dollars in funding to start building toward that vision. Once the funding was unlocked, we started helping Kolmeo build its team.

It was so much fun to speak with our contractors about working with Kolmeo. The opportunity to work with them really sold itself.

Scott nailed the product vision. His story was electric. It connected. It moved and excited everyone I spoke to.

A Persuasive Vision Boosts Recruitment

I've had the opportunity to work with many exciting businesses, both well-established as well as those that were scaling. I've also worked with many dull businesses. I've witnessed high performers being attracted to these companies, and I've seen these same high performers being repelled. In most cases, the difference was the company's vision.

High Performers Crave Transformational Opportunities

There are so many jobs out there. In this post-pandemic job market, it is an employee's pick of the jobs they want. Many organizations are struggling to keep up with filling positions in their company. The best performers in the market won't just work for anyone. They must know that the role is going to provide a challenge. And not because of the drama and dysfunction in the company. High performers are looking for positions that promise to take their careers from zero to 60.

A transformational opportunity is one where the high performer will leave your organization and have companies falling over themselves to recruit them. Transformational opportunities are the ones where the candidate cannot wait to wake up and come into your office, or these days, log in to their computer from home. Transformational opportunities are the jobs that employees find it difficult to log off from at night because they are learning and growing while working for you. A strong vision creates these opportunities and delivers a massive amount of healthy energy to your workplace. Your big vision creates transformational opportunities.

High Performers Want to Make a Difference

Right now, we're seeing a huge shift in the job market. We're seeing so many candidates run away from investment banks and high-priced consulting companies and flock to not-for-profit organizations and non-governmental organizations. We're seeing some of the best and brightest minds step back from careers that are just geared toward making lots of money and leap into those that have a clear connection to making a difference in the world. While your vision may not be about saving the planet or ending homelessness, it must be crafted to show how you are impacting people in a positive way.

Let's be clear. We're not talking about corporate social responsibility. So many companies have picked up pet initiatives because they know it attracts and retains high performers. A vision that breathes life into your recruitment efforts is one that is clear on how you're making a difference in the lives of your customers and in the lives of your immediate stakeholders.

If your product really doesn't make a difference in the lives of your customers and stakeholders, there are still ways to make your business meaningful, but they must come from your heart. You must choose something that you truly care about and make genuine efforts to contribute your time and resources to that cause. You must also make this initiative a core part of your vision and not just an add-on. If you haven't done this in your company from the beginning, it's better to do it sooner rather than later.

Creating an Energizing Product Vision

Before you go hot and heavy into recruiting and building a team in your scaling business, ensure that you're leading the product. The following are several checks you can use to ensure your product vision is energizing. These checkpoints can be developed into a survey so that you can gather objective data and revise your vision as necessary.

Your Vision Is Clear and Simple

A clear vision can be explained within five minutes. It can create intrigue and interest. I have talked to some leaders for over an hour about their product vision. Often, a core vision is there beneath all the words, but it's often not clear and I have to search for it. If people must ask a lot of questions to clarify exactly what your vision is, something is wrong. Test it out. Find a new person and talk to them about your product vision. See how long you need to explain your vision before they can present it back to you succinctly.

Your Vision Is Communicated Repeatedly

An energizing product vision is communicated regularly. While the vision may be clear, people simply forget. Under the fog of the day-to-day grind, it's easy to forget what our North Star is. Your job is to hold that vision. To create space and opportunities to communicate it. And then to communicate it in different ways and at different times. One of John Kotter's eight reasons for transformation failure is under-communicating the vision by a factor of ten. Kotter writes,

> "I've seen three patterns with respect to communication, all very common. In the first, a group actually does develop a pretty good transformation vision and then proceeds to communicate it by holding a single meeting or sending out a single communication. Having used about .0001 percent of the yearly intracompany communication, the group is startled that few people seem to understand the new approach. In the second pattern, the head of the organization spends a considerable amount of time making speeches to employee groups, but most people still don't get it (not surprising, since vision captures only .0005 percent of the total yearly communication). In the third pattern, much more effort goes into newsletters and speeches, but some very visible senior executives still behave in ways that are antithetical to the vision. The net result is that cynicism among the troops goes up, while belief in the communication goes down." (Kotter, 2007)

Kotter goes on to tell leaders who want to communicate well that they should include messages in their hour-by-hour activities. They must communicate how a proposed solution fits into the bigger picture and how an employee's behavior either helps or undermines the company's vision. When questioned on a company's objectives, they should be able to tie their answers back to goals that ladder up to realizing the vision.

Additionally, leaders must use all existing communication channels to broadcast the vision. The company communications are often packed full of boring information, but the vision should always be lively and interesting. Leaders with an energetic vision can turn routine management meetings into exciting discussions about progress toward that vision.

Your People Should See Their Place in Your Vision

It was always disheartening to me when someone on my team admitted that they couldn't see how they fitted into my vision. It means we had either hired badly or that we hadn't been clear in our vision. One of my startups was developing a new learning platform. In preparation for implementing the product with our corporate customers, we hired a seasoned corporate trainer to join our team. Most of that team was composed of product designers, learning designers, and other digital specialists. Unfortunately, our recruit didn't grasp the vision, which, in hindsight, should have been a warning sign. Ultimately, and not surprisingly, that product was not successful. Your people should know where they fit within your organization as well as how they fit into your company's vision.

Your Vision Excites

An energizing vision should generate excitement. Now, don't expect a Steve Balmer dancing on the stage level of excitement about your company's vision. But you know your vision is exciting if your people feel proud to talk about the great work they are doing and what your company is going to achieve. When a vision is exciting, you rarely hear people talk about their jobs or their work without referencing the customers they are serving. Your team should be keeping your vision—the creation of a new and better world for your customers—at the forefront of their daily discussions.

Your Vision Is Divisive

On the surface, this sounds negative. But a vision that doesn't divide is likely not clear or ambitious enough. Energetic visions have the power to turn people away. They force people to self-select. Some visions can even cause anger or generate doubt in your abilities to bring the vision to life. Those are all good signs that your vision has energy.

Your Vision Unites

On the other hand, an energizing vision connects the right people. It makes it clear who your partners should be and what company objectives to focus on. It helps your team to organize themselves because they can see the pathway to obtaining the vision.

The Story of Your Product Must Align with *You*

Once you've created an energizing product vision, you need to align yourself to it. An attractive and powerful vision must be *embodied.* You must live it. You must be able to hold the space and weight of that vision. What do I mean by that? I mean that everyone in the organization will be looking to your leadership. Every process within the company is going to ladder back to what you say about the company's vision. Every move that you make needs to be in line with your vision. What you do speaks much louder than what you say. The stories that you tell and the stories that people tell about you should link your past experiences, successes, and failures to the position that you hold today.

When new employees join small and growing companies, it's unlikely that you've established the brand yet. There's only you—the person who represents what the company is and what the company can become.

If your story doesn't match your vision, you'll struggle to find congruence in your leadership and your people will notice. A lack of congruence will undermine your efforts to bring on talent as it'll reflect in your product. Additionally, attracting and retaining the right talent becomes incredibly difficult when you're not genuinely excited about that vision.

Your Product Vision Should Allow You to Let Go

When you see the importance of vision and how it relates to people, it's a wonder how you could find much time to do anything else. In fact, the truth is that you shouldn't be doing much else. I was the COO of a scaling business where the CTO held the product vision, but he massively undermined that vision by staying too close to the work rather than holding the vision for his team. He would jump into the development environment over the weekend to code new product features. He would change his team's code without consulting them. When they came back in on a Monday, they would look at all the work he had checked in and quickly and quietly begin to undo it.

For Patrick, all he'd known was coding, but he missed the importance of leading the product. He had recruited very talented and capable developers, but rather than making his vision clear and trusting his team, he became obsessed with checking over their work and ensuring things were developed the way that he wanted.

Many entrepreneurs find that once their businesses reach between 15 and 30 people, they can no longer keep track of everything that's happening. Fifteen is the number of people that it takes to no longer have good oversight of what everyone is doing. Meaningful communication begins to fracture. It's at this stage that clarifying your product vision is even more important. Patrick's involvement in coding and designing solutions while trying to hold the product vision meant that both jobs were done badly.

Some quick takeaways from this chapter:

- A clear, meaningful, and powerful vision really matters for top performers.
- Strong visions attract and repel. If your vision does neither, you will attract aimless and mediocre talent.
- Your job is to keep everyone focused on your vision. That requires you to get out of the details.

Knowing where you are going requires you to continually take stock of where you currently are—of your position, your strengths, your weaknesses, and your uniqueness. Now that we're confident we have a clear picture of your vision, let's use that information to build the precise amount of company culture necessary to help you reach your North Star.

› REFERENCES

Kotter, John P. "Leading Change Why Transformation Efforts Fail." *Harvard Business Review*, January 2007. Accessed February 3, 2023. https://www.lifelongfaith.com/uploads/5/1/6/4/5164069/___leading_change_-_kotter.pdf

3

Build (Just Enough) Culture

Trying too hard, too soon with culture can backfire.

Provide enough cultural "happiness" as required and nothing more. Creating belonging, connection, and purpose at the workplace is the game we play to get people to stay at our company and work harder.

You don't need to look too far to understand the power of culture. A brief look at the world's most valuable company will tell you all you need to know about the value of building a strong cultural identity.

A *New York Times* piece detailed the Apple Store's staffing practices and its ability to achieve upwards of $500,000 in sales from employees making $25,000 a year. Apple achieved this by creating a cult-like culture.

Apple's applicant pool consists of disciples—fanboys and fangirls who are already in love with their products. One former employee stated, "You've always been an evangelist for Apple and now you can get paid for it." (Segal, 2012) According to journalist David Segal, new recruits are greeted with standing ovations and told repeatedly that their mission is to "enrich" people's lives. "The idea is to instill in employees the notion that they are doing something far grander than just selling or fixing products … [Apple] understands that a lot of people will forgo money if they have a sense of higher purpose." (Segal, 2012)

A company culture that can create an almost religious fervor (and ridiculous margins) is something that Apple has worked for decades to create and maintain. While building such a strong culture may be a good long-term strategy, for your company to advance through its scaling phase, your ambitions will need to be much more realistic.

Stop Trying So Hard: The Law of Reversed Effort

Have you ever tried to pet a cat but every time you approach it, the cat runs away and stares at you from a distance? Eagerly, you walk toward the cat a second time, but it runs away again. Finally, the cat flees and disappears when you approach it for the third time. A few hours later, the cat walks toward you and jumps on your lap. You'd forgotten all about the cat because you focused your attention on something else.

This phenomenon is called the law of reversed effort and was coined by Aldous Huxley. He said, "The harder we try with the conscious will to do something, the less we shall succeed." I have also found this principle to be true in life. The more that you want something, the more difficult it becomes to obtain that thing. The idea is to let go, let things unfold naturally, and eventually, you'll find yourself having achieved the thing that you wanted.

A unique and distinct company culture can truly set your company apart within your industry. You can create both raving fans and raving employees who help you generate outsized margins. However, for rapidly scaling organizations, it's better to allow most of that culture to "come to you."

Striking the Right Balance

I want to present you with a new definition of company culture. This definition is for you—the CEO or founding member of a scaling business. Culture is your customers' and employees' lived experience of every single decision you make. So, quite literally, culture is everything you choose to do. Now, wait a minute Reg! Didn't you just say, in a previous chapter, that culture wasn't everything? Correct, and this is the paradox of culture in a scaling company.

Culture isn't everything for successfully scaling a business. But culture is realized in everything. Culture is whom you select for your founding team. It is the city in which you choose to establish your business. Culture is the color of your branding and the words you use to describe your company. Culture is what time you and your team begin your workday. Culture can even be created by how you and your team arrive to work. Do people regularly cycle into the office? Do they take the train or drive? Culture is the restaurants you choose to go to for lunch. It is whether you use Microsoft 365 or Google Workspace. It is whether you offer new recruits a Mac or a PC. Culture is the good behavior that you praise and the poor behavior that you overlook. Culture is in the way you celebrate and in the way you punish. I think you get the point.

If you want to attract and retain great talent in your scaling business, you can neither be overly concerned about culture nor totally unconcerned about culture. You must strive for an ever-present awareness of the culture you're creating. That means you're aware of both the positive and negative impacts of your decisions on culture, but not overly concerned. This is because it is notoriously hard to scale culture. The startup culture you fostered when you were a tight-knit group of 20 will be obliterated once you begin to grow quickly. Therefore, many of your early hires must be flexible to the cultural volatility of a scaling business. Ideally, they would thrive in that environment.

Figure 3.1 Attract and retain talent through cultural awareness

So, how do you strike that balance? Let's look at three approaches to finding it.

Maintaining Cultural Awareness

Cultural awareness is the discipline of actively managing the impact of all your decisions on company culture. Many of your most impactful decisions have already been made, such as your choice of industry and the types of customers you'll serve. However, you should take care to maintain cultural awareness when hiring executive leaders and people leaders. Besides your industry, product, and customers, these key hires will have the biggest impact on company culture. You'll really want to consider the cost-benefit of achieving your goals and the positive and negative impacts on the culture that each hire will bring.

Even seemingly small decisions can have an impact. Be aware of how you'll be perceived when you decide to skimp on the two-ply toilet paper to save a few bucks. Yes, that really happened. More important than macro-level decisions, is maintaining awareness of your and your leadership team's day-to-day behavior toward others. These small interactions will have the greatest impact on culture.

Monitoring Leadership Behaviors

When people show up to your office in the morning and you greet them and ask about their weekend, you convey care and belonging. I've worked with entrepreneurs that were so busy and focused on their goals that they ignored most of the people on their teams. Surprisingly, even saying hello, a very small gesture, has a significant impact on culture.

A behavior with a more consequential impact on culture is how you treat people exiting your company. If people exit your business and you fail to acknowledge them with some sort of public goodbye or announcement, others will perceive you as cold and uncaring. They will likely lose respect for you. Other simple practices include occasionally going for lunch with your team. This shows that you value who they are individually and that they are worth your time.

You must remember that your actions should be aligned with your goals. If you know that you'll need to replace a significant percentage of your team at your next big milestone, you may choose not to invest the effort today.

Investing Too Much, Too Soon in Culture

In many fast-growth situations, entrepreneurs and leaders may end up overdoing it with their focus on culture. This is particularly true of founders who have been intimately involved in their companies for many years before scaling. It's also true of leaders who have had to make too many allowances to retain key employees while seeking product-market fit. If you've become overly concerned about culture, you risk creating unrealistic expectations among your team. It is possible to give too much. This is usually to the detriment of your long-term business goals and possibly even your own health.

Culture building is time-consuming and very challenging to undo. The worst-case scenario with overdoing it is that your employees begin to expect too much. You may inadvertently create a sense of entitlement. I've seen this happen many times. And I've also personally made this mistake.

When I was building one of my businesses, I made it my mission to create a company where people felt welcomed, felt they belonged, and that they could be themselves. My desire was to develop a culture where women were genuinely equal to men and knew that the men in the businesses supported them and their ascendance into leadership roles. I wanted people to feel as if they were valued members of the team and that they each had a voice that mattered—because they did matter. Unfortunately, that business struggled for a long time because I created a sense of entitlement among that team. I was overly concerned about the culture.

The truth about a scaling business is that there are going to be so many decisions you're forced to make that will compromise the culture you're trying to create. For instance, you might value loyalty, but you also need to promote a newer candidate because they are a better performer. Or, perhaps your employees are demanding that your leadership team listen and engage them more; however, the pressure from your investors doesn't allow much time for feedback.

There's no doubt that you'll find it incredibly difficult to find the right balance between growth and developing a productive and pleasing culture that attracts and retains the best people. If you overpromise, your eventual inability to deliver on your promises will cause people to lose faith in your leadership. It is generally better not to make promises and then, when they are least expecting it, you can strategically surprise and delight your employees.

I have also been on the other side of the spectrum, led by people who don't give enough or care enough about their employees. And I have consulted with a lot of businesses with these "unconcerned leaders." Those are not fun places to work. These leaders are often focused on the results with little regard for the *how* or the cost to culture. In the worst cases, this creates a hostile environment where people are bullied into submission. These places are easy to spot. Employees hate their jobs; they step on one another and are solely concerned with their own interests. Employee retention is non-existent, and jobs at those companies are like a revolving door. The "unconcerned leaders" approach may provide short-term benefits, but it often significantly undermines long-term results.

Minimum Viable Culture

Landing somewhere in the middle toward cultural awareness will help you to develop a "minimum viable culture." This means that the attractiveness of the culture you create should only marginally outpace the acceptable limits of what's required to successfully attract and retain the best people in your company. This approach is pragmatic and cost-effective. You won't have a perfect company culture, and you're unlikely to be deserving of a "best place to work" award, but then again, is *any* company, really? Aiming to develop a culture that lands just above the acceptable limits for your best talent is the most beneficial course of action for a scaling business, for a few reasons.

Firstly, it forces you to reflect on the stage of your business and the type of talent you need for that stage. We'll talk more about this topic later in the book, but for now, just remember that getting the right people for where your business is *today* requires a minimum standard of culture that will allow you to attract and retain those people.

The second benefit of aiming to build a minimum viable culture is that it creates enough "space" or "room to grow" by not hardening your cultural practices until the absolute last minute. This is critical because your business has not settled into itself yet. You're likely to experience more change, requiring you to rethink your cultural practices.

Finally, this approach is sensible because it keeps costs down. As a management consultant, I was involved with many cultural initiatives. These programs are extremely expensive and tend to distract from the core mission of the organization. Suddenly, everyone is focused on hygiene factors rather than growth.

However, if your culture is below the minimum viable limit, I hate to tell you, but you will need to do some work to improve things. Not maintaining that acceptable minimum limit means you will struggle to attract and retain the type of people that you need to take your scaling venture to the next level. However, there is an additional solution that also carries a price tag. It's called "burn and churn," and it's exactly what it sounds like.

With the burn and churn method, you simply accept that your company culture has serious deficiencies and that you have no intention of addressing the issues (for now), thus forcing faster attrition cycles.

Let's examine both strategies.

Cultural Deposits and Withdrawals

Improving the culture sounds nice in theory, but it often means changing out key leadership roles. Later in the book, we'll look at strategies for maintaining alignment among leadership. At this stage, you just need to recognize that improving culture will come at a high cost. Remember, culture is the cumulative effect of all the decisions you make.

A helpful way to think about culture is by comparing it to your bank account. When you've got a large balance, you're more likely to feel happy, confident, and secure. When you begin making debits against that account, you are cashing in and receiving some tangible benefit from the resources you've built. At the same time, you're lowering your bank balance and reducing future confidence and security. In the worst-case scenario, your account goes negative. Again, you've gotten some benefit from the resources spent, but now you're in the red. Too long in the red and you're in trouble.

I had been brought in as the COO for a scaling technology business because I had prior experience leading product delivery teams and developing product support functions. When I started, we had a tech support team made up of highly experienced senior software developers. It was certainly not the type of work that senior software developers expected to be doing when they joined our innovative scaling business. I knew we needed to create a proper technology support team with an adjacent customer success team. There existed a massive opportunity here for us to better connect with our customers, provide a better customer experience, and create an area of the business where we could onboard new talent at a low cost and then promote them throughout the business.

As happens in every business, I found myself dragged from one fire and into the next. Although those senior software developers had been rotating through the support function, it really bothered a couple of the guys who had been promised much sexier work when they joined our company. The decision to postpone developing that support team was a cultural bank withdrawal. I ultimately didn't lose any staff because of the decision, but it was a breach of trust that we struggled to recover from. From that point on, I tried to make more deposits with those senior software developers than withdrawals.

I've come to learn that your team will mostly be happy working for you so long as you make more cultural deposits than withdrawals. The goal here is to maintain a healthy cultural account balance. There's no need to keep massive cultural reserves until you've strategically determined why you'd want to maintain a high balance. Some of those reasons might include retention, engagement, or attracting talent.

Accepting a High Churn Rate: Burn, Baby, Burn!

Scaling businesses that are pursuing a burn and churn strategy are generally making more withdrawals than deposits. New recruits into the business who were once excited to work there slowly become disenfranchised with the organization and leave. This strategy increases your turnover rate of employees and can carry negative consequences if the talent market perceives your business as one that burns out people quickly. However, depending on what business benefits you're buying with your cultural withdrawals, it may make perfectly good sense to do so. Scaling businesses with the goal of being acquired may choose to follow this strategy. Similarly, when you're operating in a winner-takes-all market, you may need to have less regard for culture.

Generally, it won't be a simple choice between minimal viable culture or burn and churn. I've seen many scaleup businesses pursue a combination of both strategies. More often, I've observed that scaling businesses have "pockets" within their companies where the culture is great and the people are having a fantastic time working there.

Recently, I was onboarding a new team member to be placed with a client. She was concerned about the culture of the organization. To assuage her fears, I asked her to speak with two contractors who were working with this client. My desire for her was to have a clear picture of what she was getting into.

The first person she spoke with told her that she absolutely loved contracting for that client. Her team was fun, the work was challenging, and she had the freedom to design products with no constraints. Her next conversation with the product owner was less than stellar. He had been reporting to a manager who clearly hated the job and made work suck for everyone reporting to him. And that's often the case. One group is having a great time, but another part of the organization feels their work is a living hell and spends quite a lot of time, after hours, searching for a new job.

Company culture can be viewed much like any other asset within your business. You must determine when to invest in it and how much. But the most important aspect here is developing the right culture at the right time for where you're heading.

Some quick takeaways from this chapter:

- Giving too much, too soon creates a culture of entitlement.
- Create a "good enough" culture, or a minimum viable culture, that allows you to reach your next milestone.
- Don't try too hard, too early. Allow your unique culture to emerge.
- Be mindful of your emerging culture. Take corrective action if things go off the rails, but if you're going for broke, you'll undoubtedly sacrifice company culture for the sake of your future success. Own it. You can always fix it later.

There's a popular military saying that *slow is smooth and smooth is fast.* In the final chapter of Part II, Refine Your North Star, we'll look at why you'll start to move more slowly as you begin to scale, and how you can use this to your advantage.

› REFERENCES

"5 Signs That Apple Is a Cult." The Week. January 8, 2015. https://theweek.com/articles/474359/5-signs-that-apple-cult.

Isaac, Mike. "Inside Uber's Aggressive, Unrestrained Workplace Culture." *The New York Times.* February 23, 2017. https://www.nytimes.com/2017/02/22/technology/uber-workplace-culture.html.

Thomson, Jonny. "The Law of Reversed Effort: The Harder You Try, the Harder You Fall." Big Think. May 14, 2022. https://bigthink.com/neuropsych/law-reversed-effort/#:~:text=The%20Law%20of%20Reversed%20Effort%20was%20first%20coined%20by%20the,the%20less%20we%20shall%20succeed.

Thomas, James. "Bullet Holes & Bias: The Story of Abraham Wald." Mcdreeamie's Meducational and Medhistorical Musings. July 8, 2019. https://mcdreeamiemusings.com/blog/2019/4/1/survivorship-bias-how-lessons-from-world-war-two-affect-clinical-research-today.

Segal, David. "Apple's Retail Army, Long on Loyalty but Short on Pay." *The New York Times.* June 23, 2012. https://www.nytimes.com/2012/06/24/business/apple-store-workers-loyal-but-short-on-pay.html?ref=business&pagewanted=all.

4

Ready, Set, Slow Down

Why overcapacity is the secret to successfully scaling your teams.

Know what it's going to cost to start something new. Then double it or perhaps triple it. Everything costs more and takes more time than you originally expected.

Philip, an executive in a large financial services company, needed to deliver marketing campaigns that were data-driven and backed by experiments involving their customers. The plan was to transform his marketing department and make it agile so it could quickly deploy tests in the market and build better marketing campaigns.

The transformation would create efficiency by delivering marketing strategy and data analytics as a service. Once in place, they would scale-up the marketing tactics and execution teams. They were working with a base of nearly 200 marketers and were expecting to bring on another 100 people. When we rolled out the new structure and trained everyone in agile ways of working, Phil's business stakeholders saw immediate benefits. Instantly, we could see bottlenecks in their processes and were able to resolve them. Communication improved significantly and productivity shot through the roof. Phil looked like a genius.

However, Phil hadn't anticipated how many new issues would surface because of that restructuring. Confusion was rampant. Many marketers were unclear on who their new managers were and which marketing projects they'd been assigned to. There was confusion about role descriptions, and there was a lot of duplication of effort.

The restructure reached a new low when they tried to scale on top of the new structure. When they began hiring and growing their teams, existing communication issues were compounded. Several marketing initiatives missed their deadlines significantly or were canceled because of delays. One of the biggest signs of failure was when people started to quit. It looked like the entire transformation might fail.

A lot of people were demanding things go back to how they'd been before the restructuring. However, Phil and his executive team held a firm position that they would not roll back the changes.

Truthfully, though, they had already begun to quietly reverse some of their biggest decisions. The executives also began to create more time for things to normalize before introducing additional changes. They were still ambitious in their goals, but they were somehow more supportive in their execution. They shifted their focus away from a rapid transformation toward more incremental improvements.

We did eventually complete the restructure, although it didn't entirely resemble the vision that was initially promoted. The result was that the marketing team did ultimately deliver more of the right strategic marketing programs by embedding experimentation. The new ways of working that they learned ultimately led to greater efficiency and better customer experience.

Growth Is Never Linear

Figure 4.1 Expectation vs. Reality

The pathway to our destination is never straightforward. And, through those difficult journeys, we're ultimately changed too. Figure 4.1 is an incredibly simple depiction of what business or personal growth often looks like in reality. We like to believe we're pretty good at going from point A to point B. It's human nature to set a goal, chart our course, and march off in confidence. We're sure that our plans are foolproof and that we'll be successful. But, much like Phil learned, growth is almost never linear.

We constantly face setbacks, disappointment, unexpected hurdles, and sometimes, downright chaos. Our plans, when subjected to the daily pressures of reality, often look half-baked, immature, and possibly even naïve. My goal in this chapter is to present you with ideas to help you blast through some common challenges you'll face as you begin to scale. I'll also present you with some focus areas to help you prepare for your next big hiring push.

Adding More People Is a Trap

Consider the orchestra with its many different musicians. Each one of them crafts different sounds and beautiful melodies to create a cohesive and masterful work of art. Now imagine the conductor doubles the size of that orchestra overnight. Would anyone expect this oversized orchestra to instantly make more beautiful music? A sensible person wouldn't expect them to make music as good as the original orchestra. A cynical person would expect that they'd gotten worse!

When you scale-up your team, people don't automatically begin working well together. They require time to normalize and adjust. You must allow time for the four phases of team development to occur: forming, storming, norming, and performing. The stages occur sequentially, and they create room for team growth and progress.

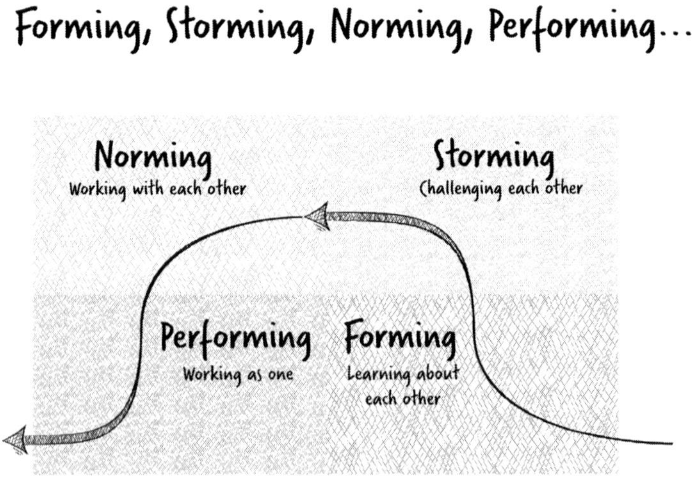

Figure 4.2 Four Phases of Team Development

The most notoriously overlooked fact about this model is that every time a team member is added or exited, the team composition is changed. This means that the team must begin its journey through each of the four phases all over again! This is particularly notable for scaling businesses. It shows us that when you begin scaling, the impacted parts of your organization are vacillating between forming and storming for a period in line with how long it takes you to scale.

Take, for example, the workload involved in hiring a new team member.

You'll likely need to write a job description, get it approved, advertise it, review the applications, and interview each applicant. Then you'll need to give them an offer, wait out their notice period from their prior job, onboard them, train them, help them acclimate to the new environment, settle in, and get comfortable. That period alone could be anywhere from six to 12 months.

Then, and only then, are we finally back to the workload we had before making the decision to hire someone new. Additionally, our new hire only starts to increase your capacity at *this* point.

Now, consider the effect of doing this with not just one person but with many, and across many teams. Consider the disruption this will create across multiple areas of the business, throwing out the balance of workflows you've worked so hard to establish. Mass hiring will also test the limits of your company culture, which can have a direct impact on productivity.

And remember, the reason you're hiring people is that there is already more work than your current team can handle.

Does this sound like a recipe for success?

We can visually demonstrate this concept in another way. The slow ramp-up curve in Figure 4.3 below shows that as we begin to scale and add new staff, productivity initially begins to flatten and dip. As time progresses, when scaling our teams, productivity does eventually rise; however, it significantly lags behind our hiring efforts to speed things up. We're now trapped in a persistent lag in productivity created by our scaling efforts.

This outcome is highly counterintuitive to our beliefs about increasing capacity through hiring, but it marries well with our observations of team members getting caught in the forming and storming phases.

This is an important concept that can be especially critical to understand when convincing investors to keep up their investment activities in your business. Adding people generally creates new problems, drags productivity, and negatively impacts culture before we start to see benefits. Despite all the sophisticated analysis and cost projections that are inevitably done, I've seen many scaled projects blow their budgets because of the productivity lag generated by hiring.

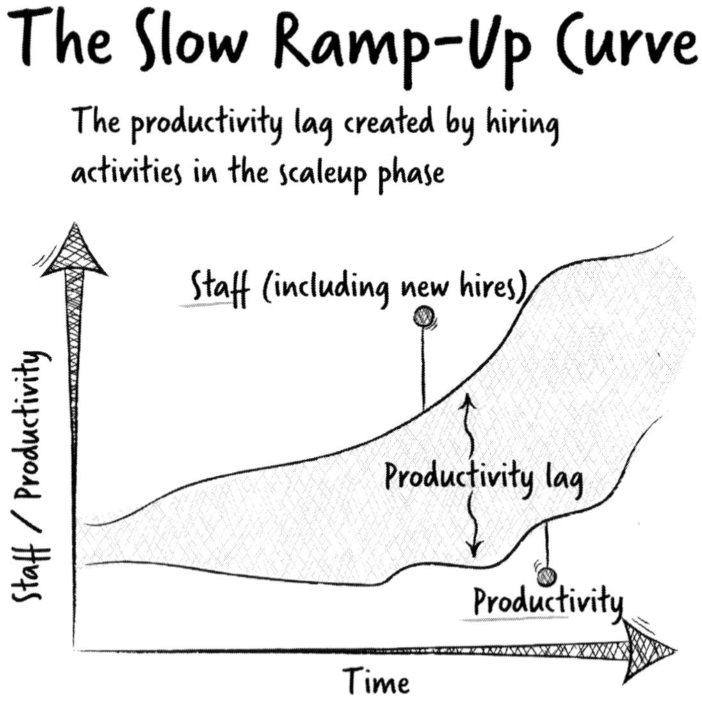

Figure 4.3 The Slow Ramp-Up Curve

Hire When You're Ahead

As we've seen, having the resources for scale does not always ensure success because of the productivity lag. However, we can overcome the productivity lag by being smart with our timing.

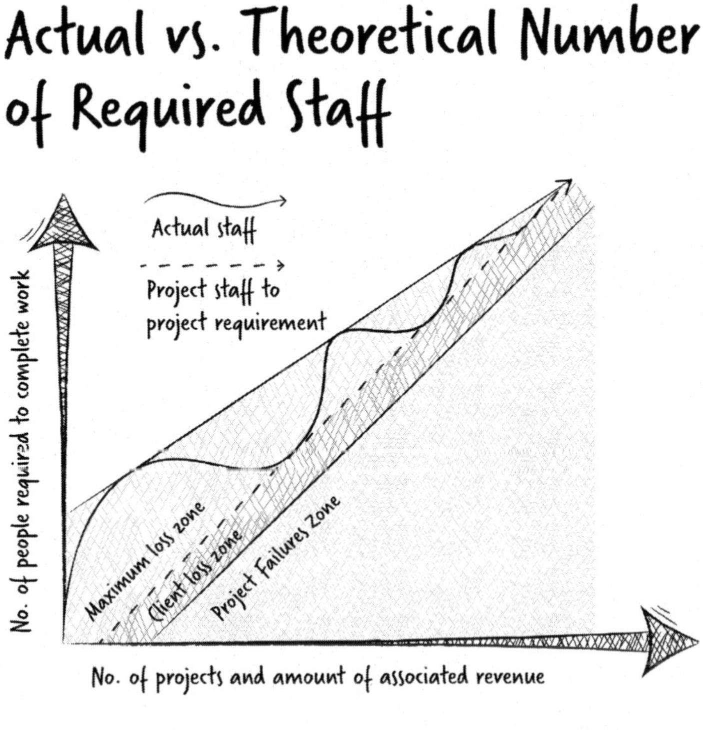

Figure 4.4 Actual vs. Theoretical Number of Required Staff

To grow quickly and overcome the productivity lag, you need to hire more staff than projected—enough people to get out of the "maximum risk zone."

The maximum risk zone is the space where you theoretically have enough people to complete all your projected work. However, due to the impact on productivity that new hires create, you're technically understaffed.

The often-overlooked solution here is to overhire.

When was the last time you went to any major sporting event or concert where they ran out of beer, hot dogs, and popcorn? Thankfully, this rarely happens! Good event organizers will forecast demand based on previous events and then oversupply the stadium. Sure, at the end of the night, there's a bit of waste. This is a small sacrifice for capturing maximum revenue and ensuring all patrons have the best possible experience. It's much better to have more resources available than you need at the beginning than having to reactively add more resources to meet demand.

There are a few key lessons that this model teaches us for beginning our scaling journey.

Firstly, stay ahead of the game. Avoid big hiring pushes when you're already behind schedule. The notion that you can "catch up" is a fallacy. The productivity lag that will be created when you begin hiring means that you'll never hit your goal in time. Your hiring activities will create so much disruption that you'll find yourself stuck in a relentless cycle of needing to hire more people to close the gap.

JC was managing a project to create a mobile supercomputer. Using the computing power of thousands of inactive phones on the network, the downloaded algorithm would process cancer research data. While novel and innovative, the client had demanded more features than we could possibly keep up with. The shortfall of resources and the pressure to deliver led to significant quality issues with the product. Rework and team burnout led to further delays. Once the client obtained more funding to increase our resources, the project experienced a productivity gap that only caused further delay.

In these situations, I recommend that you prioritize relentlessly and cancel any unnecessary plans. Invest your resources in a place that you have time to properly prepare for.

The second key takeaway from the event model is that you should make your case to investors for additional headcount as soon as you get wind of a major opportunity. Doing so will give you the extra capacity required to pass the maximum risk zone and will be your best chance of minimizing the disruption created by the productivity lag.

Minimize the Productivity Lag

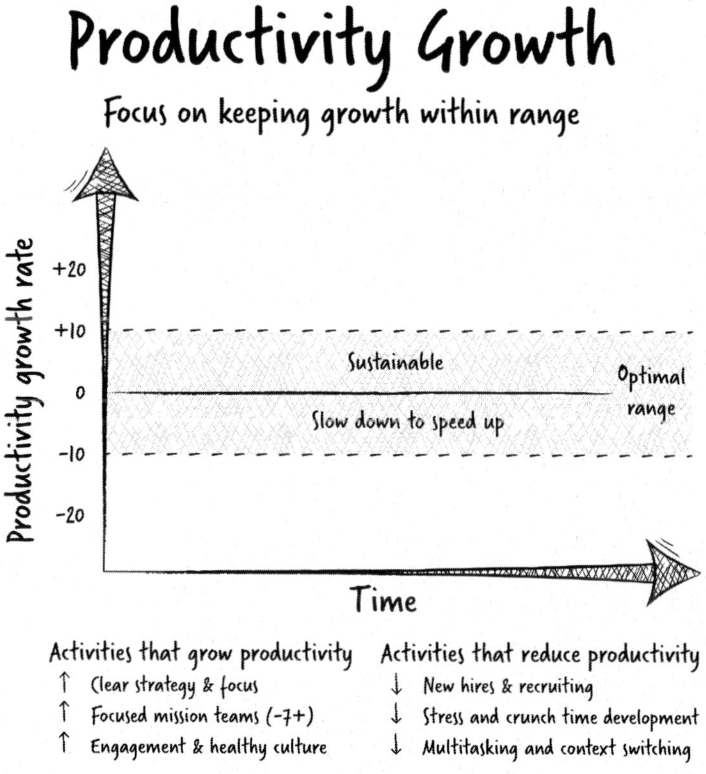

Figure 4.5 Productivity Growth

The productivity lag you'll experience when scaling cannot be avoided. However, Figure 4.5 presents a simple tool to help you manage growth without destroying productivity. Some activities that promote healthy culture and productivity include:

- A clear strategy and focus.
- Focused mission teams.
- Clear priorities.
- Regular team engagement.
- Steady investment in improving culture.

Investment in these activities will help you lead through periods of high growth. The following is a list of necessary activities that are detractors and work against productivity.

- Recruiting.
- Team involvement in interviews.
- Onboarding new hires.
- Crunch time in hitting project milestones.
- Multitasking.
- Context switching.

While these activities will create unavoidable disruption, you can create balance and minimize productivity lag. Using this productivity growth framework, you can easily monitor whether or not your plans to scale will be sustainable.

The other option for minimizing the productivity lag is to simply go around it. My client was a scaling technology company with a mix of projects and various products they were developing. They found themselves caught in the "catch-up" fallacy. They believed that continually adding new resources would help them speed up development and ship to their customers sooner. What really happened was that their continual hiring activities burnt the team out and made productivity worse. That was until they discovered Readify.

Readify was a specialist software development agency with developers who had been working together for years. Working like a specialist team of navy SEALs, they were mission oriented, self managed, highly motivated, and extremely tactical. Each time a new team was onboarded from Readify, we made significant progress in developing the product. We ring-fenced these software developers from much of the internal red tape. This allowed them to produce two or perhaps almost three times as much as the existing software development teams. Hiring entire teams of people with this level of experience can be your ticket around the productivity lag.

Sharpen the Axe as You Scale

In addition to minimizing the productivity lag, you'll want to prepare more holistically for growth. Here are some useful ideas to consider as you refine your North Star in preparation for scaling.

Reset your lens — If you're looking at people as chess pieces that can be moved around or reconfigured without adverse effects, you might want to adjust your thinking. Every time you alter the configuration of your teams, onboard new members, or make any other substantive change, you'll impact productivity.

Find your pace — Talk to your teams to understand what they believe is their capacity for training and integrating new hires while still delivering on work. Use that information as a baseline for defining your sustainable hiring pace. Finding your pace will also help you understand how much you can push the limits without causing harm.

Plan for downtime — Scaling works best when you move through periods of hard work and then periods of rest. Periods of rest don't need to be long. They can be as little as two days each quarter. A growing company is like a living being. After periods of hard work and physical exertion, your body needs sleep to restore and repair itself. Although it may seem that these two days of rest is time wasted, allowing your organization the time to heal and reset itself will create the space and energy to sprint again.

Reset expectations — Plan for your normal work rhythms to require adjustment and refinement as your team grows. As we'll examine in later chapters, communication problems abound as your team grows. Tension will arise when old delivery cadences remain the same with the burden of hiring added on top.

Improve communication — Run regular surveys internally to get feedback on how well you're doing with communication. One of the keys to success and sustainable scaling is ensuring everyone stays on the same page.

Update processes and systems — No matter how solid your systems and processes are, you'll start to see them hit their limits when you begin to scale. Start documenting which ones you believe are going to break when you start hiring. Develop plans now to enhance them while scaling.

Get ahead of the curve — When and where feasible, overhire. If you've discovered there's great talent in the market due to a competitor downsizing, hire that talent now! Doing so can help you to get out of the maximum risk zone and ease tensions when project demands rise.

Design a flexible structure— Design an organizational structure that is flexible and scalable. As you scale and onboard new leaders, you'll need to redesign how you're organized until you find a structure that supports your current stage.

Reduce your risk — Consider hiring existing teams who have already worked together before. Set up partner relationships with service providers who provide specialist contractors. These people are professionals at getting up to speed quickly and producing fast. We'll explore these solutions in more detail in Chapter 8.
Organize for success — Putting people into small, mission-driven teams helps to maintain agility. Consider the impact of a small

team slowing down versus an entire department or team.

Some quick takeaways from this chapter:

- As you scale, expect things to slow down before they speed up.
- Ensure you have the budget to overhire so that you can get out of the maximum risk zone.
- Avoid big hiring pushes when you're already behind schedule.
- Leverage your resource suppliers. The teams of people they provide can get you around the productivity lag.

Now that you've fully refined your North Star, you're ready to begin your hiring push. Next, let's explore more deeply who to hire. We'll go through different strategies you can use to gather your crew. To begin, we'll examine your most difficult job as the CEO of a scaling business: building a team with the *right people.*

PART III

GATHER YOUR CREW

Part III of The Scaleup Culture Cycle is the largest because it is the most critical component to your scaleup's success. We cover this area quite extensively to prepare you for building a winning team and developing a culture that will support you in achieving your vision.

We'll look at two team-building methods available to you for scaling your team: one well-known and one commonly overlooked, yet powerful technique. We'll also explore how you can improve and mature your team-building processes over time. Finally, we close by discussing a key method for maintaining alignment as you grow.

The culture that is ultimately cemented into your business will take shape based on whom you choose to begin scaling with. So, we'll begin Part III by examining your hardest but single most important job: getting the right people.

5

Get the Right People

Building a team with the right people is your hardest but single most important job.

Branding is powerful. People will believe what you say about yourself. Just make sure you can deliver.

He is incredibly charming, funny, and smart. He can talk to anyone. He's motivated and he's ambitious. You're excited about him joining your team. You have massive expectations about him transforming a part of your business, really shaking things up, and getting things moving. You know this guy is going to make waves at your firm, and he's excited to join, too. It looks like a match made in heaven.

As it goes, a few months down the track, you realize he's a dud.

Go ahead. Insert their name here ______. I know you have at least one story like this of a new and promising recruit who turned out to be a massive disappointment. And, if you haven't experienced this, you won't have to wait long. Eventually, it'll happen to you. Extra points if they come with a massive ego!

Building a team with the *right* people is your most difficult but single most important job.

Why Hiring Is So Challenging

First off, if you're reading this book, it's likely that you're in a scaling business. Or it's likely you're *hoping* to be in a scaling business. And you know what? That's great! You're embarking on the ride of a lifetime! But I'm almost certain that you've never done this before. Even if you're like me and have been involved in multiple startups and multiple scaling businesses, the particulars of your service, product, or platform are different. The people you're working with are different. This is a brand-new experience for you. You have never designed and built seats for this particular bus. In every company, the roles are going to be a little bit different. In every stage of your scaling business, the roles are going to shift dramatically. Unfortunately, this creates a situation that's primed for failure. And there are many failures to be had.

One reason getting great talent in the door can be such a difficult proposition is that you're moving so quickly. You've got a clear mandate to build your vision, you've got the cash, and now it's ready, set, GO! Moving at breakneck speed increases the likelihood of making more disappointing hires. A 2019 study from Leadership IQ found that 46% of new employees will fail within their first 18 months of employment. The key reasons for failure included lack of coachability, low levels of emotional intelligence, poor motivation, the wrong personality, and technical incompetence. (Wharton, 2022) And these statistics reflect normal working conditions, not the increased pressure of a scaling business. Do you think you can beat the odds?

Another reason it's so challenging to get the right people into your scaling company is that, let's face it, how you advertised the role is quite often not anywhere near how your company really functions. Chances are you're selling an aspirational version of the business, not the true one. If you told prospective hires what was really going on in your company, it's unlikely that you'd ever be able to convince anyone to join!

This means there's often a huge disconnect for the person you're bringing into the company. They are often not prepared for the realities of a new and scaling company. They may not know how to navigate the complexity of a business that isn't fully formed and is changing so rapidly. Finding people who are talented and capable but who are also not turned off by volatility and uncertainty makes your job even more challenging.

The world is changing dramatically. People are exiting the workforce at breakneck speed. The rise of digital technology has made it possible for people to freelance anywhere around the globe. While writing this book, I've seen how the coronavirus has upended how people work everywhere. The current trending term is "quiet quitting," where employees are doing the bare minimum in their jobs. The worst-case scenario is that you bring on one of these people who work on their side hustle while they are meant to be working for you. Currently, one of the most egregious trends is people holding down two and sometimes three full-time jobs while working remotely.

Finally, you're not exactly clear on where you're going as a company, and this makes predicting who you'll need in the future very difficult. Of course, you have a rough idea of what needs to happen. But, as we've said before, this is your first time doing this. You might think you know what you need to accomplish today, but in six to 12 months' time, that reality will likely be different. Once you learn that, it'll be too late, and you'll be stuck with people in your company who are looking more and more like they don't fit.

The Wrong Hire

A few wrong hires can set your scaling business back a few years. An investment in the wrong senior person, for example, can have a disastrous effect on your business. One of my clients brought someone on with the responsibility of leading the development of a new suite of products. The individual was extremely well-known in the industry that they were trying to disrupt. Just having his name on the website gave the business great credibility. But it turns out he didn't have such a great grasp on the market. His

$260,000 annual salary was a complete waste, but much worse than the expense, that one bad hire set their product schedule back by two years.

It takes a lot of time to advertise, recruit, interview, and hire an employee to replace someone. For example, the cost of replacing an executive-level employee is, on average, 213% of the annual salary for the position, according to an article by Mason Group. The cost of a bad hire is estimated to be 1.5 to 3.5 times their annual salary. The real cost to your organization for hiring the wrong product manager earning $160,000 annually, therefore, will range between $240,000 and $560,000. And this only refers to costs that you can measure and calculate objectively, such as the cost of posting, the cost of HR time, the cost of managers involved in the interview process, and the cost of training. But this is just the tip of the iceberg. It doesn't take into account the less tangible costs of, for example, the reduction in productivity, the effect on retention if employees are unhappy, and the opportunities lost that could have been capitalized on by the right hire.

Besides the assets on your balance sheet, your company will never be more than the sum of the talent in the business. If you're lacking critical capability, it's because you have not yet developed those skills in your people, or you have not yet brought it in. Your business is like a collective brain when you bring all those talented people together. If you're going to soar to new heights and reach your next major milestone, you'll have to find the best hires available in the market and maximize the capabilities you have in-house.

HR Can Only Get You So Far

Does all this sound too daunting for you? Are you thinking: Why am I reading this book? Why don't I just hire an HR person? Or perhaps you already have an HR person. Problem solved!

Unfortunately, you cannot simply outsource these talent sourcing issues to an HR person and think that that'll be the end of your problems.

The HR person, as well as the external talent sourcing people you've brought on, can only run the mechanics of a people and culture department—things like developing policies, developing standard operating procedures, implementing your employee value proposition, executing a search process, etc. But, at the end of the day, the type of people that enter your organization to build your vision will be selected by you and your leadership team. You have the final say on who makes the cut. And a key component of your role will be your ability to match your talent needs to the stage of growth your business is currently in.

Matching Your Team to Your Stage of Growth

Your business is less like a machine and much more like a living organism. Think of a newborn baby. It needs constant attention, nurture, and care. But that baby's needs will change as they reach the toddler stage. They'll begin exploring and require slightly less attention. They'll seek freedom in a safe and controlled environment. Then consider its childhood, adolescence, adulthood, and senior years. Each stage has different requirements. Your business is much the same. Each stage requires different types of people. Let's look at the most well-known stages of a business and explore the people that are needed. And let's unpack how your business will require those people to change to meet the next stage, or rather, how you might need to change the people.

Startup

The first stage of a business's life is the riskiest. There are many reasons why businesses fail, but according to the U.S. Bureau of Labor Statistics, only 80% of startups with employees survive the first year. In the startup phase, you're likely spending most of your time and effort bringing your business idea to life. You are probably trying to promote your product or service while also managing other important tasks. You're probably wearing many different hats, which is common for startups trying to get their business up and running.

So, if you're currently a small business, working on your product, and looking for product-market fit, those 20 or so people on your team are likely the best team to get you there. They are agile and, like you, they share roles and switch hats a lot. They happily pick up tasks from each other and they are excited about building something completely new. However, that entire dynamic changes once you reach the growth stage.

Growth

Congratulations, your company has been rapidly growing since it started a few years ago. The business is working. Your product or service is known to consumers. Your revenue is increasing. There is less turnover as your market share and customer base grow.

Now you've determined your next major funding milestone will take you to a team size of about 50 people. That's incredible, and it's going to be an exciting adventure. But are those 20 original people going to be the best choices for the next stage of this journey? How about your journey from 50 to 100? Or 100 to 300? What generally happens is that your business requires more systems and processes to support scale. This means more predictability and repetitive work. Some people crave this sort of environment. They need routine and want clear outcomes. But many employees of startups find this stifling and, at this stage, move on to new startup businesses.

You'll need to have a plan in place to help those employees either grow into more senior roles or move on to the next opportunity. We'll discuss this later in the book. However, one way that a lot of scaleup product companies navigate these early growth complexities is through leveraging contract-to-hire positions. Contract-to-hire allows you to scale your team up and down as demands change.

The benefit here is essentially being able to try before you buy. You can bring in some high-quality talent, people who can hit the ground running. Contract-to-perm allows you to weed out many candidates who would not fit your business in the long-term. It also allows you to attract practitioners who are at the top of their field. These contractors are often highly engaged as well.

Maturity

You're likely feeling safe and secure when you reach this stage of the business life cycle. It's a different experience from the first two stages. Because you didn't have an established product or service, the startup phase was risky. In the growth stage, you have the pressure of managing *how* your business grows so it still accomplishes its goals.

At this stage, you're starting to convert much of your temporary staff over to permanent ones. You're focusing much more on managing your costs to streamline your most profitable services. As a business that has been around for a while, you won't be sitting still. Your company has the chance to expand. You can increase your market penetration by ensuring a larger percentage of customers are using your product or service. You may want to develop new products to tap into a new market. You might also be thinking about selling, merging, or buying another company to expand during the maturity stage.

You likely now have multiple types of employees, from juniors all the way to senior and late careers. It's also highly likely that, during your growth stage, you've developed the right people and cultural processes to manage the growth and development of your staff.

Renewal or Decline

It can be difficult to tell if your business is declining. As your customer base grows, you may find yourself meeting higher demands. If your business has seen several years of dropped revenue, you're in a decline. It is important to regularly look at your finances so you're aware of your spending.

At this stage of the life cycle, you may be thinking about reinventing your business. If you reinvest in your company, it could be renewed. If you want your business to avoid decline, it's best to start this process sooner rather than later. If you notice a change in the industry, modify your strategy. This could very well mean that you need to work with new people. We'll explore this topic in more detail in the next chapter. For now, understand that you can introduce new thinking into your business by working with consultants and contractors and converting those roles to permanent ones.

Designing the Right Seats

Your ability to get the right people into your scaling company is a function of how well you design your seats. And the design of the seats in your company rests completely on your business strategy. While strategy is outside the scope of this book, I will present you with some practical tips to ensure you have the right setup for success. Additionally, I'll outline a simple test you can use to get a sense of whether you've designed the right seats.

Know What You're Looking For

Before you run out and hire, make sure you have some clear position descriptions available about the job you have on offer. It's a simple suggestion, but you'd be surprised how many growing businesses have missed this point. It might be a good idea to reach out to some HR consultants and benchmark certain roles.

Have a clear understanding of what the market has to offer and what caliber of person you're going to need for your current and future stages of business. Expect that you won't get it right the first time around. This is common and you'll need to factor that into budgeting. One of my clients hired three separate people into a role before eventually getting it right.

Utilize contractors and consultants. Oftentimes, these people will design the job role for you. Having had significant prior work experience and exposure to many different types of organizations, these individuals often perform at a higher level and will provide a template of what "good" looks like. They are also more likely to better navigate the complexities of a scaling business.

Trust, but Verify

Accountability, for many leaders, is attempting to "catch" employees doing something wrong, pointing the finger at co-workers, or creating a strict set of rules administered with a punitive approach. This negative approach drives a reactive culture of "management by rules" rather than creating a proactive atmosphere of responsibility. But there is a better way.

Setting and holding people to a common expectation by clearly defining the company's mission, values, and goals is what accountability in the workplace is all about. It is when every employee is responsible for achieving business goals, from the part-time hourly worker to the C-suite executive. Verne Harnish has a useful definition of accountability in his book, *Scaling Up*: "The rule: if more than one person is accountable, then no one is accountable, and that's when things fall through the cracks."

When It's Right, It Feels Good

It's typically clear you've hired the right people when the work of that team is both challenging and fun. When you have the right people for the right stage of business, you're truly enjoying the process. Something electric and magical happens when you see your ideas coming to life, but the people who are making it happen are also enjoying their work, enjoying each other, working hard, and growing.

I once led a team that had a mixture of people from various, ethnically diverse backgrounds and with extremely different capabilities. These team members had very disparate ways of seeing the world. Duong, having escaped Vietnam as a child during the war, believed our team could overcome anything if we stuck together. That didn't always mix well with Priya's desire to stand out. A brilliant solution architect from India, who grew up in a competitive family, she always wanted to take the lead and be the best. The tension was sometimes palpable. But that team had a blast creating and delivering our product. It was one of the most fun teams I ever led, and we accomplished so much and got so much work done. That team was so successful because I was very deliberate about every decision I made about whom I brought in. I thought very clearly about the positive and negative attributes of each person. Having the right mix of skill and tension within a team is important for bringing out the best in people.

When you've got the right team, something special happens. People become dedicated to one another. That's why it's so important for you to make smart choices. That dedication to one another creates a trench-warfare mentality. When you're standing shoulder to shoulder every day, fighting for the same outcomes, it creates such an amazing feeling of camaraderie and partnership. These types of bonds are impossible to manufacture through cultural initiatives. This type of culture is born by getting the right people on board and giving them a mission that aligns with their skills and desires.

Of course, not every team is going to be fun and challenging. Expect to have pockets of people who drag themselves into work. They churn out code or business requirements, and they hate it. You should expect the attitude that they show up to work with will reflect in the quality of the product that they create. Not only that, but their attitudes *will* creep into your culture. Make sure you keep people like this away from mission-critical assignments.

When your team is having fun, enjoying their work, and feeling challenged, it makes your job much easier. Or at least it'll look much easier from the outside. There are always going to be people issues to deal with. These are people, after all. But the job starts to look easier. Your team becomes much more self-organizing, taking initiative, and being accountable. When they are humming together this way, it's not time to take your foot off the pedal. You still need to be vigilant and make sure they stay excited about the work.

Once it's clear you've onboarded the right people, you must keep selling them the job. They may still need to be convinced that they made a great decision. Making sure you do this early can keep your best people with you for years to come. In essence, you're re-recruiting them. In fact, you're recruiting them for the next role in your business. Many popular job sites recommend that employees stay with a company anywhere from two to four years. Depending on market trends, I have often seen this time drop as low as 18 months. You also need to keep yourself informed about the market. In Chapter 8: Refresh Your Talent, I'll introduce some tools you can use to help you know when your people are getting itchy feet and should move on.

Some quick takeaways from this chapter:

- Face it: When you are hiring at scale, it's difficult. You're going to get many of those hires wrong. Build it into your cost estimates.
- Even if you create a talent-sourcing function, getting the right people is ultimately your key responsibility.
- The type of people you need will change drastically as you cycle through phases of growth and maturity.
- You'll know it's working when your team starts creating deep bonds with each other. You should be concerned if the atmosphere never feels right.

Bringing great talent into your scaling organization will be challenging. It'll be a combination of finding people with the right skills and the willingness to ride out the ups and downs of volatility with you. You'll need to be prepared to design and redesign job roles because no one ever gets it right the first time. And you'll need to get good at matching talent to the stage of your business. Many people will be able to help you navigate this journey, but it probably won't be your current talent recruiter.

› REFERENCES

"The Real Cost of a Bad Hire." The Mason Group. September 13, 2018. https://www.masongroup.ca/employers/real-cost-bad-hire/.

Wharton, S. Thomas. "The 5 Reasons New Hires Fail in the First 18 Months." Miles LeHane. Accessed December 16, 2022. https://www.mileslehane.com/blog/the-5-reasons-new-hires-fail-in-the-first-18-months.

"Why Is Accountability Important in the Workplace?" PowerDMS. Accessed December 16, 2022. https://www.powerdms.com/policy-learning-center/why-is-accountability-important-in-the-workplace.

6

Fire Your Recruiter

How the traditional recruiting process is broken and what to do about it.

Revenue is vanity.
Your number of employees is vanity.

Back when I was leading product delivery teams in scaling businesses, I was approached by a recruiter. She called me and told me about a fantastic opportunity with a really impressive new scaling technology company.

"What type of platform are they developing on?" I asked eagerly.

"Oh, it's the most scalable platform available on the market," she mumbled, trailing off.

Probing a bit deeper, I asked, "Have they brought in a head of product, or do they have a CTO leading the product function still?"

"Umm, the CTO is finalizing some critical technology decisions about the product. You'll be there to lead the team." Her answer made very little sense to me. I thought I'd try again to see just how well-organized their delivery capability was.

"What organizational methodologies are they leveraging to scale the development of their product teams?"

"Methodologies?" she asked in a concerned voice. "You know, I'm sorry. This is only my first week on the job. I'm going to be honest, I don't know anything about the company or the job. Why don't I find out more and then call you back?"

Unsurprisingly, this is one of the more positive experiences I've had with a recruiter.

The overwhelming majority of recruiters are fast-talking salespeople who often don't have an ounce of experience in your industry. In this chapter, we'll explore how the traditional recruiting process is broken, and I'll provide some strategies for getting the best talent into your scaling team.

Buyer Beware: Dreadful Practices of the Recruitment Industry

There's certainly a massive need for recruiters. The market for hiring and recruitment in America is worth a whopping $28.5 billion dollars, and it's growing at an average of 5.3% each year. The average sunk cost of a new hire is somewhere between $4,325 (Deloitte, 2015) and $41,111 (Oxford Economics, 2014). When you account for the (unreported) time spent by in-house teams searching for candidates, that number can increase substantially. What we find, however, is that an overwhelming number of recruiter's placements fizzle out within the first year or shortly after. The average annual turnover rate is 57% across all industries, which accounts for both voluntary and involuntary turnover, according to a 2021 study by the Bureau of Labor Statistics. However, the actual rate of turnover varies greatly by industry. According to a 2019 study, tech companies are at serious risk, with an average turnover rate of 20.9%, the fourth highest overall behind retail, manufacturing, and consumer goods.

A key contributor to much of this turnover is poor candidate placement on the part of recruiters. Many recruiters are simply spouting jargon-filled gibberish to the candidate and then blatant nonsense back to the hiring manager about the candidate. The technology landscape can be complex, and the market can shift quickly. This, combined with an abundance of recruiters who have never worked in the industry, is a key cause for much of this behavior.

Recruiters who do not have experience in the industry face difficulty in asking the right questions of the candidate and of the hiring manager. Many recruiters fail to offer true value to their customers because they do not possess the knowledge to do a deeper level of candidate screening. What generally happens is that the hiring manager ends up reviewing many resumes that are not the right match. Additionally, the recruiter often passes over many candidates that are actually a good fit.

One of the biggest issues with recruiters is the transactional nature of their work. This business model, based on a flat commission on placement of a new hire and found in most traditional recruitment companies, can be harmful because it means they don't provide a true partnership and they're driven by a short-term goal. The recruiter takes the brief about your job vacancy, conducts a search, and presents you with what you're told are the best candidates for the role. What may not be entirely clear to you is how this "cash-on-placement" model tends to drive a range of behaviors that can damage your business in the long run. Let's further explore some of those practices and then unpack how they each harm your interests.

Overselling the Candidate

Jackson was looking to move up in his career. He had been a digital project manager in a marketing technology agency and felt that he'd hit a ceiling. He was looking for a new challenge. Having read an interesting job advertisement about a program management role for a startup company backed by one of the leading banks, he contacted the recruiter to discuss the position. The recruiter loved Jackson's resume. He offered him some information about the company.

"You'll have access to the latest industry tools, and you'll have the freedom to work on the projects that you want," the recruiter advised. He continued, "The team is advanced, with cutting-edge technology. They are leaders in their field."

Jackson was excited. He progressed through the interview process and was offered the role, which he accepted. But within six months, Jackson was searching for a new job. The opportunities that had been promised to him didn't exist. Jackson was still seeking a role that would fulfill his career ambitions.

Candidate overselling is one of the most common behaviors running rampant throughout the recruitment industry. Oftentimes these claims are untrue but said with confidence to encourage the candidate to take the interview. The underlying motivation is to shape the candidate's perspective of the role so that there is a match.

Many recruiters will also exaggerate the growth potential of your company. Individuals are looking for a ride in a fast-growth scaling company, particularly in the startup/scaleup space. Recruiters will therefore play up the company's technology and ability to disrupt in their industry. Although you may have those ambitions for your business, recruiters must leverage those stories correctly and not oversell the true potential of the company.

Minimizing Your Challenges

Transactional recruitment business models also drive the recruiter to undersell the challenges that your company is facing. Rather than painting a clear picture of your company, your unique challenges, the state of the business, your growth cycle, and the internal issues that you're dealing with, recruiters tend to paint a rosy picture. I get it; recruiters need to create excitement because this excitement fuels the candidate's desire and increases the chance of setting up an interview and inching themselves closer to the transaction.

You might be led to believe that some of these behaviors could be of benefit to you because it means you'll have a full pipeline of talented people interested in working for you. Unfortunately, there is a high cost to a sales approach that exaggerates the benefits of your company. It creates downstream issues when the candidate is ultimately disappointed with where the company is once they come on board.

I've witnessed many disappointed employees say, "I expected this successful technology company to have had a better onboarding process, but it wasn't very good." Or, "I thought these guys were using agile processes for delivery, but they are running a full waterfall project management framework." The candidates' dissonance between expectation and reality sows the seeds of disappointment and leads to early attrition. And the truth is that this loss begins during the recruitment process. But your recruiter will have already moved on before it becomes apparent. You'll be left paying the costs of that loss long after.

Exaggerating the Candidate's Abilities

If you think it's bad practice to oversell the candidate, would you be shocked to know that your recruiter is also using this tactic on you?

Recruiters will find a candidate with an impressive resume and begin to talk about their experience and abilities. Their job is to get you excited about their potential. Once you're sold on the candidate, they use scarcity sales tactics to seal the deal. They dial up the pressure by emphasizing the risk of losing the candidate. They will say things like, "This candidate will not stay on the market long," and, "This applicant already has other offers, so you should make an offer now." At the same time, they will tell the applicant that this job opportunity won't last very long.

While it may be in your interests to act quickly, many recruiters will use this high-pressure situation (that they themselves have created) to get the applicant to request a higher salary. The purpose of having the candidate ask for a higher salary is to effectively increase their own commission. The result is that the recruiter has created the perception of fit to successfully increase their margins on the transaction.

When the placement eventually goes wrong, you won't have

time to think about what went wrong and whose fault it was. Instead, you'll be back in the market looking to replace the role because you have work to do.

Candidate Recycling

A truly insidious and manipulative practice from transactional recruiters is what I call "candidate recycling." This is when recruiters poach the same people they placed less than 12 months ago, in order to place them in a different firm to generate another transaction. This is the single most unethical practice facing traditional recruitment companies.

I've personally had recruiters who placed candidates in my companies reach out to my new employees within six to 12 months in the guise of checking in. What I later learned was that these recruiters were floating new job opportunities with them, coming from some of their other clients. This practice is extremely damaging to your bottom line, and you'd do well to stay away from any recruiters who are known to use it.

In-Group Bias

A final and common problem I've seen with recruiters is in-group bias. This is the tendency to favor someone from our own group. Let's face it: We're human and we tend to gravitate toward people like ourselves. The issue, however, when we don't seek diversity of thought, is that we create groupthink and generate blind spots.

Recruiters often build talent pools from the same organizations and with people who share their group dynamics rather than evaluating candidates using different sources and from different backgrounds. On the surface, searching for candidates in proven places might seem like good business practice. The problem is that, over time, your organization will become filled with people who share the same beliefs and thought patterns. This is proven to lead to lower levels of innovation.

Aligning Incentives through Full-Service Recruitment

The reality is, if you're leading a fast-growing technology company, you're going to have to work with recruiters. I've given you a picture of some of the worst practices in the recruitment industry. Now the question is, how do you select a partner to work with and ensure that you avoid those practices? The answer is to find a full-service recruitment agency with a business model that demonstrates that they have skin in the game. They will provide value above and beyond simple recruitment. As a growing and scaling business, you're going to need to develop a range of processes to ensure a steady stream of talent at the right levels. A full-service recruitment agency can help you. In your search for a full-service recruitment agency, the following sections outline the additional services they should provide.

Developing an Employer Value Proposition

The author of *Employer Brand Experience*, Brett Minchington, defines an employer value proposition (EVP) as: "A set of associations and offerings provided by an organization in return for the skills, capabilities, and experiences an employee brings to the organization." It's the total value of your company to the potential employee, including the salary, benefits, training opportunities, the company's brand value, and the quality of the leadership. It's an important consideration for candidates when they are looking to join a new company, but it's also a useful way to retain current employees. It's therefore important to spend time and effort developing a compelling EVP. This should be something that a full-service recruitment agency can help you with in order to attract the best talent. But it must be aligned with existing, integrated workforce planning strategies, so the agency should know your business intimately. It should also be informed by the experience of existing employees as well as the external audience that you are targeting.

Developing Your Talent Acquisition Processes

A full-service recruitment agency should also offer services that allow you to build out your talent acquisition process. Recruiters and talent management are very similar and they overlap; however, there are some significant differences.

A company's short-term headcount needs are addressed by traditional recruiters, much like the ones described at the beginning of this chapter. Talent acquisition, on the other hand, which is what a full-service recruitment agency will provide, is an overall business and HR strategy that factors in an organization's long-term goals and acknowledges that people (or talent) can play a huge role in a company's future success. Talent acquisition is an ongoing process set up to identify and vet appropriate candidates for executive-level positions, leadership roles, and jobs that require specialized training, rather than simply filling seats.

Developing Your Onboarding and Offboarding

Later in the book, we'll talk in more detail about onboarding. Really, you're looking for an agency that can help you develop processes for ensuring that as people come into your company, they are set up for success. Offboarding is a similar process. How people exit your company matters. You'll need to ensure that the timing of these events is handled well and in a sustainable way.

Embedding Your Full-Service Agency

Firstly, a full-service recruitment agency should be embedded into your company. How can you ensure they will deliver you a steady stream of talent that matches your organization if they don't know it inside out? Your ability to attract and hire the right talent is a strategic skill that will enable you to realize your goals. Finding a full-service partner who can embed their people into your team therefore allows them to identify talent that is culturally aligned and at the right level to support your business growth over the long-term.

A Skin-In-The-Game Business Model

A full-service recruitment agency will have a business model that is win-win-win. The business model of a recruitment agency should never allow them to win while you or the job candidate loses. A simple way to ensure this is the payment of commissions over a period of time, such as a year. In exchange for offering the recruitment agency some exclusivity on certain roles, or promises about how many roles you expect to bring on, they will accept payment of commission over the long-term. Another model is to look for an agency that will replace the role at no extra charge if the candidate decides to leave within the first year.

Some quick takeaways from this chapter:

- Transactional (cash-on-placement) recruitment business models are notoriously abused, and you should steer clear.
- Find a recruitment service that shares the downside risks with your business.
- Look for a full-service recruitment agency that provides long-term talent acquisition services.

Recruiters can be a tricky group to work with, and finding an honest one can be a huge challenge. But you're in luck. Recruitment services aren't the *only* way to get amazing talent into your business. In the next chapter, let's see how you can position yourself to take advantage of serendipity by procuring some underutilized sources of talent.

› **REFERENCES**

Quilty, David. "What Is Employee Value Proposition (EVP)?" Firstup. December 7, 2022. https://firstup.io/blog/what-is-evp/.

7

The Serendipity Factor

How risk, luck, time, and chance play an outsized role in building your team.

The pursuit of your desired future is more satisfying than achieving it.

My family works, lives, plays, and invests across the United States, Canada, and Australia. Australia is one of my favorite places because of both its natural and created beauty. The entire world is familiar with Australia's stunning beaches. And it would be hard to imagine Sydney without the iconic silhouette of the Opera House. The world-renowned building has a background full of challenges that its designers and engineers had to overcome for it to exist today.

An international design competition was launched in 1955, calling for a venue suitable for operas, ballets, and concerts. Two hundred and thirty-three entries were received from 32 countries. The Opera House was to be built in three stages: first, the concrete base, then the curved "shell" roof, and finally, the interior and theatre machinery.

Construction began in 1959, but the designs had not actually been finished at this point. This raised questions about the building's structural integrity. In fact, the foundations had to be rebuilt because of inadequate work early in the construction, as the roof's weight couldn't be supported.

The Sydney Opera House's design was centered around the "shell" roof, but the engineers who were tasked with constructing the concrete panels struggled to find an acceptable solution. At least 12 different options were ruled out by the design team.

The original cost estimate was $7 million, and it was scheduled to be completed by January 1963. By 1973, it was finally complete after a 14-year wait, costing a grand total of $102 million. It was ten years late and 1357% over budget.

The story of the Sydney Opera House is an example of the planning fallacy. This is our tendency to underestimate the amount of time it will take to complete a task, as well as the costs and risks associated it—even if it contradicts our experiences. This fallacy can cause us to plan poorly for the future by making decisions that ignore realistic estimates of the demands of a task. An overly optimistic assessment of our abilities leads us to downplay the effects of risk and luck.

To take advantage of risk and luck, I never discount the "serendipity factor." Serendipity is the phenomenon of finding good things in life that we haven't searched for. In life, much of what occurs happens by chance. My career began in Philadelphia rather than Washington, D.C. because the IBM talent manager sent my job offer to the wrong email. I found myself living in Australia and working with startup and scaleup businesses because I randomly had an Australian manager who invited me on secondment to Australia. I've made lifelong friends who were witnesses at my wedding because I said yes to a random invitation to a Kenya gala.

So much of life and business we try to plan out and control—but I have found that much of what we do and where we end up is largely outside of our control. Our job is often to simply make the best of the opportunities that we've created or that have been presented to us. In this way, we can overcome the planning fallacy and build all-star teams by recognizing the "serendipity factor" at play in much of what we do.

Latent Serendipity in Hiring

Strategy is important, but don't underestimate time and chance. Some of the most exciting parts of running a business are the good things that happen that you could've never planned or predicted. Many of the best and most qualified job candidates are already gainfully employed in someone else's company. So, bringing them into your business will require you to source them using a variety of techniques. Let's examine a few and see how serendipity is already at play.

Headhunting is a popular method of finding and sourcing talent. Headhunters will help you craft a clear position description that you're trying to fill. You will need to provide them with a crystal-clear picture of the type of person you want to hire. Let's say you're looking for a chief marketing officer. You might desire someone who has worked in your particular industry, in Fortune 500 companies, with a specific track record of transforming brands and entering overseas markets. If you find a good headhunter, they will help you narrow down the qualities of that person even further using additional demographics like age, sex, ambitions, etc. Your headhunter, who is likely already familiar with your industry, will also craft a company profile for you, advertising the strengths and selling points of the company. With that information, they will move forward and conduct their search.

The next step is to generate leads and begin a process of converting early candidates to prospective employees who are willing to sit and have a conversation with you. The process can take anywhere from six weeks to six months depending on what you're looking for and what role you have on offer. I've seen some roles take as long as a year to fill. These services are reserved for the most specific and high-value roles in your company. Expect to pay a handsome sum for this service.

You can get the most from these services by recognizing that your specific search criteria may be difficult to fulfil. You'll likely need to make compromises on your wish list, and this is where the exciting element of serendipity comes in. Your search will probably return a range of candidates who have selling points that you were not looking for and are possibly unaware that you might have even needed. Some of these candidates could cause you to redesign your role and maybe even some of your business objectives if they stand out as exceptional. You can easily spot exceptional candidates, as they tend to start adding value to your business through the interview process by means of advice and consulting. As you go through the interview process, don't simply focus on the role you're looking for, but seek to learn more about how the role could be crafted or shaped around the individuals that have been sourced in your search process.

For the majority of your team-building process, you'll likely use a typical recruiter. Most recruiters' approach to job search is to maintain a database of candidates. Good recruiters should regularly speak with talent in the marketplace to understand if they are happy in their job, getting itchy feet, or very much in the market for a new role. From that pool of candidates, they will advertise your staffing needs. They, too, will follow a structured process for screening candidates, shortlisting them, and putting the best talent in front of you.

In both scenarios, serendipity is at play but not being maximized. Time and chance are enormous factors in why the most qualified candidate for your job may be willing to move versus being uninterested in hearing about your job at all. Who eventually takes your available position and whether they will truly fit your organization, have the skills required to do the job, and have the right attitude to grow with you, is not guaranteed by either of these methods.

Take Advantage of Serendipity

Taking advantage of serendipity isn't a management process but rather a philosophy for life and business. It requires you to remain in a state of expectation and positivity. As an entrepreneur, you've no doubt already learned to be an optimist. You'll need to channel that same energy into your hiring practices—not necessarily in the robustness of the hiring processes that you set up, but in your willingness to experiment and step outside the box when it comes to building the right team.

Taking advantage of serendipity is particularly useful in young, scaling businesses. As you'll see in the coming sections, the methods for taking advantage of serendipity will help you onboard an extremely diverse group of highly skilled people. Rather than building all your capability from scratch, you'll be able to buy it in and then convert these people to full-time employment. In the growth stage of business, this is beneficial because it creates the opportunity for diverse thinking and experiences in ways of working while still moving toward your goals.

The first out-of-the-box method for building a team that takes advantage of serendipity is called contract-to-permanent. I first introduced this in Chapter 5. As I said before, I consider this the "try before you buy" method. Contract-to-perm means you allow the contractor to prove themselves before you make a long-term decision to bring them into your team permanently. This is a tremendous benefit to you because many experienced job candidates today are incredibly skilled at selling themselves, yet they simply do not perform well over the long-term. The contract-to-perm method allows you to onboard highly skilled and experienced professionals quickly, see them in action, and gauge their fit before deciding that you'd like them to join you. Often, you find they are skilled operators, but they're perhaps not the best long-term fit for your company. In that case, you would simply conclude the relationship at the end of the contract. Let's examine what makes this method superior to the traditional transactional recruitment model.

Exceptional Experience

Generally, contractors have significant experience from having worked with a wide variety of companies. They have seen the good, the bad, and the ugly in their field and they bring a wealth of experience that can be difficult to procure elsewhere. A key benefit of having some of these people on your team is that they bring a clear picture of what best practice is. They generally know what tactics work and what don't. They will come in, get up to speed quickly, and execute efficiently, having performed in these roles many times before.

A clear benefit for a scaling business is the contractors' experience with proven frameworks for building capability. I once worked with a scaling agency that was suffering from poor project delivery capability. To quickly build this up, they hired a team of seasoned project managers who quickly created a new project delivery and governance structure. Once complete, several of those contractors joined the organization full-time to continue leading projects and further develop the company's expertise in project delivery.

Thrive in Challenging Environments

Contractors are seasoned professionals and are often looking for the next challenging environment to leverage their expertise and hone their skills. Without moving on to increasingly complex problems, they can quickly become stale. Because of this, they tend to operate differently and charge a premium for their services because they have developed a playbook to get results. They are eager to get in quickly and start proving their worth. They are adept at making themselves invaluable to you, which means you can expect a significant return on your investment in terms of key deliverables being completed efficiently, the overall capability of your team rising, and the collective knowledge of your organization growing substantially.

Skin in the Game

The contractor model naturally creates a situation where the contractor has more skin in the game than an employee. Contractors do charge a premium, and as such, you would expect that they deliver quick and high-quality results. Additionally, this model typically provides you, the buyer, with the option to quickly terminate the contract if you do not get the expected performance. You're often able to terminate with no penalty simply because you no longer require their services. This puts you at a major advantage because you can scale-up your workforce using contractors and quickly scale down if your business needs change. Your risks therefore decrease substantially when you work with contractors.

Because your contractors are aware of these factors, they are more likely to hit the ground running. They have made big bets on themselves and are generally eager to show their value by working beyond the capacity and effort level of your full-time workforce. This is a benefit to you in multiple ways. Firstly, it means you're accelerating your projects. Secondly, it means you're demonstrating the abilities of more capable people to your current team, which creates a virtuous cycle.

The Hidden Secret

Matt was an experienced business analyst who had worked with another innovative scaleup tech firm in Australia. I had been working with a scaling property management client, and one of my business partners knew Matt was looking for a new challenge. Matt knew that most of our clients were developing innovative tech products, and he was eager to come aboard. I introduced my client, Kolmeo, to Matt. They immediately recognized him as a seasoned business analyst who was eager to work with a company that was disrupting the property management market in Australia. My client quickly fell in love with Matt, and over the six-month contract period, they were able to ensure he fitted in well with their team and continually delivered a high-quality product.

Although Matt was interested in doing more contract work with my company, he felt that he'd really found his place at Kolmeo. He had joined a team with a great culture that was looking for someone with his capabilities, his personality, and his ability to help the development team clearly understand the needs of the business. Matt decided to go permanent, and within six months, he had been promoted to the role of product manager. This is a typical story when it comes to contract-to-perm. It was a winning situation for all parties involved.

Perhaps the biggest secret to gaining talent is that you can convert contractors into permanent employees. Some companies even offer a contract-to-perm service. It's certainly advantageous to use this lesser-known technique for finding great employees. The serendipity factor states that you're likely to find good things without looking for them, and we know that much success in recruitment comes down to timing and chance. By converting those contract hires that you already know are a perfect fit, you can spend less time looking for that elusive, ideal candidate who may never come along.

So many great candidates are willing to join your organization as contractors and help your business scale. More often than not, they show themselves to be high quality and they possess specialized talent. They prove themselves capable of integrating into your teams quickly and further developing your culture. You need to be thinking, from the very beginning, about how you'll convince the best of these contractors to join you permanently.

Working with a specialized contract-to-perm provider can help you become very successful in using this method. They will work with you to understand your organization and your company culture and provide a ready-to-go network of highly qualified professionals that they are already working with and can vouch for. The work of convincing contractors to join your organization permanently will be on you, but your specialist contract-to-perm provider should be able to guide you on how to make this successful.

I've done this very thing in one of the scaleup technology companies I was a part of. My entire team came on as contractors, and I slowly worked to convert at least half of them to permanent employees, creating a highly talented team from nothing! My current company, Beaker & Flint, helps numerous clients with this very approach.

You can even recruit high-performing consultants. I have first-hand experience with this. While building and scaling an innovation consultancy, I had to constantly fight against my clients trying to recruit my people. My consultants became invaluable to the clients, and so they made every effort to bring them in-house. A lot of these companies were unsuccessful in recruiting our consultants, however, because they hadn't gone into it planning to poach them. Had they been strategic about it, I'm certain they would have been able to convert many of them by crafting a role for them in their organization from the beginning. They could have been much more thoughtful about the onboarding of those consultants by integrating them into their team more closely and by connecting with them on an executive level.

Although most consulting companies hate it when their clients do this, it is a valid way to source talent that's worth exploring, and your scaling business may rely on it. It is in your interests to poach your consultants because consultants generate an alarming amount of high-value IP that will inevitably leave the business when they roll on to a new assignment.

It's worth noting that many consultants have very little allegiance to their consulting company and are looking for a sense of belonging elsewhere. Many consultancies do a terrible job of creating a feeling of connection to the business. Look for consultancies that embed talent into your organization in an augmentation capacity rather than as a methodology-driven solution. These consultants are often forgotten about in their consulting firm and create much stronger connections to you and your organization.

A lot of consultants are also burnt out from hopping to and from client engagements. The very thing that seemed exciting to them about consulting eventually turns them off—the constant flights from state to state and the juggling of multiple clients. Most people are not career consultants. Of the top consulting firms in America, the average career span is three to five years. Most of these consultants go on to work for one of their previous clients, and that could be you.

The great thing about consultants is that, because they have had so much exposure and experience within a short period, landing a former consultant with three years of experience is as good as landing a similarly experienced person with five years of experience. These talented professionals are often substantially underpaid, making them very attractive hires for your scaling company.

Some quick takeaways from this chapter:

- Although it's common to attribute success to our own brilliance, much of it really comes down to timing and chance.
- Bringing in contractors and consultants from a service provider is one of the quickest ways to build capability.
- Leverage serendipity by onboarding contractors and consultants and converting them to permanent positions.
- Look for supplier relationships that welcome you poaching their people. If they don't advertise it, just ask them.

Like many things in life, nothing is certain. Our best and most well-thought-out plans don't stand a chance against reality. Luck and chance can and will strike at any moment, either working in our favor or against us. We have much less control than we believe we do.

As you're scaling your company and building your team, you can and should take advantage of serendipity. Allow yourself to see and experience a range of talent from places you may not have considered. The serendipity factor is a useful mindset, but eventually, you'll need to develop more long-term talent acquisition processes.

› REFERENCES

"The High, High Cost of Recruiting Developers – Dice Insights." Accessed December 16, 2022. https://insights.dice.com/employer-resource-center/high-cost-recruiting-developers/.

"Industry Market Research, Reports, and Statistics." IBISWorld. Accessed December 16, 2022. https://www.ibisworld.com/industry-statistics/market-size/employment-recruiting-agencies-united-states/.

"Planning Fallacy." The Decision Lab. Accessed December 16, 2022. https://thedecisionlab.com/biases/planning-fallacy.

"The Surprising Story of the Sydney Opera House's Construction – Google Arts & Culture." Google. Accessed December 16, 2022. https://artsandculture.google.com/story/the-surprising-story-o f-the-sydney-opera-house-39-s-construction/CAVhiDXG4On-xA?hl=en.

"The True Costs of Employee Turnover." Built In. Accessed December 16, 2022. https://builtin.com/recruiting/cost-of-turnover.

8

Increasing Your Odds of Getting It Right

Sharpen your talent acquisition process with proven techniques for building an all-star team.

Strategy can provide a false sense of security. Nothing is certain.

Travis, the CEO of a scaling technology agency, had long done all his own hiring. From the early days of his startup agency, he wrote every job description, advertised online himself, screened applicants, shortlisted, and scheduled all his own interviews. It was a lot of work. But this was his baby, and he was determined to build a dream team with the right people. Soon, his technology agency was taking off. Demand was growing, and he couldn't keep up. Moreover, he couldn't find enough qualified people.

Travis started using a local recruitment agency that specialized in supplying talent. But he quickly became dissatisfied with paying large fees for candidates who seemed to fizzle out much sooner than he believed they should. He couldn't understand why he was paying large fees to recruitment agencies for people whom he felt he could attract and recruit on his own. As the CEO of a scaling business, he chose to go back to running the entire recruiting process himself to keep his costs low and to have more control over who he hired.

Solutions for Building an All-Star Team

While I'm a huge proponent of serendipity in team building, you mustn't rely on it alone. Building the right team doesn't happen by accident. It takes deliberate and concerted effort. As your organization grows, you'll need to implement a range of processes to ensure you're bringing the right people on. The following section outlines a few key tactics and tools that I believe you, as the CEO, should lead and remain actively involved in.

Executive Involvement

Although controlling the entire process was costly, slow, and short-sighted, Travis, our CEO from the introduction, was on the right track. I've said this before, but it's worth reiterating. Building the best team possible is *your* responsibility. Whether you're using external talent sourcing agencies or you've already built an internal team, it's up to you to ensure your vision is reached. That means you need to be actively involved in the recruitment process.

For a moment, imagine being a junior software engineer who has spent two years working for a large technology consultancy. You were lucky to score this job out of college, and over the previous two years, you've reported directly to a senior development consultant. This development consultant has given you your work assignments, reviewed all your coding and documentation, provided you with notes, and worked with you to improve your core capabilities. The people you speak with daily are like you. They are other junior software engineers and software engineers who work in a similar environment. There are many senior people around you, such as the head of consulting, the technical director, and the lead architect. These are people that you've never spoken with. You are rarely in their presence. Now, imagine that junior software engineer in relation to the CEO of that company. He has only ever seen videos of the CEO. He's only ever read his words in an e-mail or on his blog. He reads about the CEO on the front cover of the *Wall Street Journal* or sees him speaking with pundits on NBC Money.

Now, imagine that junior software engineer decides to leave his fancy consulting job for your scaleup business. He's told that his final interview will be with your CEO. What does this communicate to our junior software engineer about the importance of the role that he's interested in? How does it affect his perception of your company and the tasks that he'll be asked to perform?

This is the exact strategy that Larry Page of Google famously used when hiring Googlers. Google had 53,600 full-time employees at the start of 2015, compared to 47,800 the year before. In one year, the company hired about 6,000 new people. According to Google's HR boss, Laszlo Bock, who writes about Google's hiring process in his book, *Work Rules!*, co-founder and CEO Larry Page approved each one of them.

Bock writes that, "More important than the feedback itself is the message from Larry to the company that hiring is taken seriously at the highest levels, and that we have a duty to continue doing a good job." (Bock, 2015) New Googlers are always delighted to learn that Larry Page, one of the company's co-founders, personally reviewed their applications.

When CEOs are involved in the hiring process, the quality of candidates increases significantly. When you factor in the initial relationship that is developed because of this first interaction, you can also expect a significant culture bump. In taking the time to interview each candidate, the CEO takes personal responsibility for each role and thus screens people more intensely.

Setting the Right Tone about the Importance of People

Another reason this practice is successful is the positive impact on the culture. The CEO engaging in the interview process sends a clear signal to the rest of the organization about the importance of hiring the right people. Many businesses are led by CEOs who don't spend any time with the front-line workers. For very large and growing businesses, CEO involvement in the interview process can be difficult, but for scaling technology companies, it is imperative that you are involved.

Deeper Screening

Additionally, your involvement in the interviews, as CEO, means you don't want a failed hire, so you naturally screen candidates more closely. You pay attention to small signals that indicate whether this person will be a good fit, culturally, for your organization. Although you may not understand all the ins and outs of their specialist capabilities, you will be able to sniff out someone who is all talk and who may not have accomplished as much in their previous roles as they are leading you to believe.

Keeping Your Talent Suppliers Accountable

Finally, CEO or CTO involvement in all hiring decisions means your third-party recruiters or your insourced recruitment team knows the importance of putting forward the right people to you. It heightens their desire to spot great candidates that are highly talented and who will also fit the culture of your organization. Ultimately, it ensures your sourcing team implements more robust search and hiring processes. They will do a better job because of this pressure. This is a significant advantage when it comes to getting the best candidates. As you're naturally going to be very busy, your team will be doing everything they can to see that your time is used efficiently and effectively. They may even insist that you do not need to be involved in the process and that you can use your time for more important work. Resist this request, as it quite often signals a hiring team that would rather not have you judge their performance in finding great candidates.

Topgrading

In a 1997 edition of *Directors and Boards* magazine, Bradford D. Smart and his son first coined the term "Topgrading." As they explain, "Topgrading simply means proactively seeking out and employing the most talented people available while redeploying (internally or externally) those of lesser ability or performance. More specifically, we view Topgrading as employing only A players."

However, it is not only about seeking A players; it is also about structuring interviews and making sure the process is rigorously targeting the right people. Bradford believes that talent is more important than skill or experience and that the most talented people will be the most valuable to your organization.

Topgrading requires a well-written job description that reflects the company culture and goals and the individual job responsibilities required. It looks beyond the standard interview questions. You can't attract the right people for the job without this critical information.

When you use Topgrading, candidates will face multiple interviewers. The purpose is to find the best people who fill the critical core competencies of intelligence, vision, leadership, drive, resourcefulness, customer focus, hiring, team building, track record/experience, integrity, and communication.

Here are some suggestions from the Smarts for how to go about the process of hiring using Topgrading.

1. Measure and improve the current hiring process: You need to look at your current methods before making changes. What works and what doesn't?
2. Create a job scorecard: This is a rigorous process to determine what you need for the position. You should do this before you begin recruiting so the scorecard reflects the company's needs rather than allowing candidates to sway hiring managers with charm.
3. Recruit candidates: Make sure you have a detailed job description.

4. Screen candidates with work history forms: These forms should include detailed questions for all candidates that cover, at the very least, compensation history (illegal now in some states and jurisdictions), boss ratings, reasons for leaving their previous jobs, their likes and dislikes in other jobs, and self-appraisal.
5. Conduct telephone or video interviews: This should be an in-depth screening that lasts about 45 minutes. This narrows down your field.
6. Do competency interviews: These focus on proficiency and behavior.
7. Conduct a Topgrading interview: Topgrading interviewing is intense and focuses on the candidate's entire career history, with questions about every position.
8. Provide feedback and coaching to interviewers: This helps you improve your process by giving each other feedback.
9. Write a summary: Each interviewer writes a summary of each candidate, which they can use for comparing them.
10. Have the candidate arrange reference calls: Topgrading theorizes that good performers have good relationships with former bosses and will enjoy this process. You then conduct the calls and categorize the candidates according to their perceived potential.
11. Coach the new hire: The process doesn't end on the candidate's first day. You have to make sure to make this a good environment and an excellent place to grow.
12. Measure hiring success annually: When you establish a baseline and regularly track recruitment metrics, you can identify opportunities to fine-tune and improve your hiring process.

Building Your Talent Sourcing Capability

Many growing organizations do opt to eventually build their own talent sourcing capability. Your approach to talent sourcing will form a key strategic competitive advantage that separates you from your competitors. Typically, businesses at the scaling stage will begin investing in internal recruitment teams. The benefits here are more well-maintained pools of talent with greater predictability of when candidates will be interested in the job. When you use internal people, you're less likely to be "sold" on quickly onboarding someone who may not be the best fit for your company. Lastly, an internal team will be able to standardize your approach to talent sourcing so that there's more foreseeable timeframes and consistent qualifications of job candidates.

Maternity and Paternity Reentry

An often-overlooked source of talent comes from experienced mothers reentering the workforce after maternity leave. Many scaling companies would not consider this a viable option. However, many women returning to the workforce make fantastic employees, as they are often eager to do work that utilizes the full range of their abilities. Companies that are downsizing and going through rounds of layoffs make great targets. As soon as you hear of companies known to have good talent that are restructuring or facing financial pressures, begin to target their employees. Most people experiencing disruption in their company are eager to be approached by larger companies who are looking to score great talent. Finally, if it's an option, you can always look at buying out smaller firms. This is often a great way to secure talent and is a tactic known to be used by well-funded scaling businesses.

Solutions to Avoid

As you can see, there are many different techniques and tools you can use to find great talent. When it comes to hiring and building your team, thinking outside of the box is always a good idea. However, there are some options you should simply avoid. Below is a brief list of the areas I've seen create more headaches than opportunities.

Outsourcing

You'll notice that we have not considered outsourcing so far in this chapter. While outsourcing is beyond the scope of this book, I consider it to be an inferior solution in a newly scaling business. The key reason is that many outsourced solutions are optimized for processes and systems of established and mature businesses.

When I've worked with growing businesses that leveraged too many outsourced suppliers, I've seen complete breakdowns in communication. The outsourced suppliers are often unable to keep up with the changing needs of their scaling client, and there's a tendency for the client to force additional requirements on the supplier without adjusting budget or time, leading to a significant reduction in quality.

Another key risk I must mention is the obvious cultural mismatch—developing country outsourcers, I'm looking at you. These require an enormous amount of management overhead and are still plagued by communication challenges. For the purposes of this book, outsourcing is not a viable solution for introducing new, highly skilled talent into your business.

Freelancers

For a scaling business, I consider freelancers to be an inferior source of talent. While it is possible to form a long-term relationship with high-quality freelancers, you'll likely spend too much time searching for and managing a freelance team. Not to mention, you'll be competing for priority among many of their other clients. On rare occasions, it is possible to convince a freelancer to join your team permanently. But if you're experiencing fast growth in your business, don't waste your time.

AI Solutions

I am a huge supporter of new technology solutions, but I am very concerned about utilizing AI in your hiring and team-building processes. These AI-driven solutions are just beginning to gain momentum, but their successful implementation is still a long way off. Many of them overpromise on their ability to predict performance and fit. More problematic are AI's known issues with introducing bias into the hiring process. One infamous case was the Amazon machine learning algorithm that decided women were not qualified for employment at Amazon. The engineers were forced to shut it down. Another incident involving an AI solution saw women targeted for 75% of cashier positions in supermarkets.

Some quick takeaways from this chapter:

- The easiest way to ensure you're getting the best team is to embed executive involvement in as many hires as possible.
- To increase your chances of success, implement the Topgrading approach, which systematically seeks out and nurtures quality hires.
- Avoid outsourced providers and freelancers like the plague. You'll be happy you didn't waste your time.

As your scaling business reaches new levels of maturity, it will be necessary for you to begin building more scalable talent sourcing capabilities. The development of these processes will create efficiency and reduce costs over the long run. They should also continue to leverage external recruitment or contractor agency relationships you've developed. It'll be extremely important that you maintain those relationships because you'll need to start refreshing your talent before you know it.

› REFERENCES

Bock, Laszlo. 2015. *Work Rules!: Insights From Inside Google That Will Transform How You Live and Lead.* New York; Boston, Twelve.

Feloni, Richard. "Why Google CEO Larry Page Personally Reviews Every Candidate the Company Hires." Business Insider. Accessed December 16, 2022. https://www.businessinsider.com/google-ceo-larry-page-on-hiring-2015-4.

Lucas, Suzanne. "What Is Topgrading Interviewing? Our Favorite Tips." Workable. August 26, 2022. https://resources.workable.com/tutorial/topgrading-interviewing#:~:text=What%20is%20topgrading%20application%3F,candidates%20for%20a%20given%20position.

"Role of the Board Topgrading Organization – Capital Solutions Bancorp." Capital Solutions Bancorp. Accessed December 16, 2022. http://capitalsolutionsbancorp.com/files/pdf/Topgrading_The_Organization.pdf.

9

Refresh Your Talent

Eventually, even your best talent will leave.
Know when it's time to let them go and how to do it.

Employees follow a natural life cycle of birth, growth, maturity, decline. Leverage it.

Every good thing ends. When I started my first company, I was aware of that reality. Fate had brought an amazing team of people together for a mission. I also knew that, someday, fate would also tear us apart. Still, nothing could've prepared me for a downturn in the business that would mean retrenching whole swaths of people so quickly. What the experience taught me is that no one is so unique to your business that they cannot be replaced. Everyone is expendable. In fact, you must begin with that reality in mind and design your business around the fact that they will all go one day.

Changing Your Team Members

The role you designed for today is not the role you'll need for tomorrow. When you start out, you've got a small band of people who are connected to each other and committed to the journey of building your product. They see the vision. They are excited about the possibility. It's an environment of learning, growth, challenge, and thrill. Everyone chips in. Everyone is pulling their weight, plus more.

I remember having a team that was constantly looking for ways to improve. We had people working outside of their role descriptions because they loved what we were doing. It was electric; an atmosphere filled with passion. And it was also an environment that would be impossible to scale if we continued that way of working. We needed to introduce standard operating procedures; we needed much more consistency. Unfortunately, standard operating procedures didn't appeal to a lot of the people on my team. But the choice was to either stagnate or change.

Of course, you'll have choices as well. When your team begins to show signs of reaching the limits of their capabilities, will you act quickly to move them on? Will you develop the necessary processes to plan well ahead of time for these events? Elevating your business to the next level requires that you design and implement scalable people processes. You'll need this infrastructure to enable continued iterations of your product and to deliver on ever-increasing customer demands. This is where the "three teams" framework comes in. It will give you a paradigm for linking one growth stage to the next.

The Three Teams Framework

In this section, we dive into the three teams framework. This is a useful conceptual model to understand where to place your focus as your team is growing. We'll begin by reviewing the framework's underlying principle. In the following sections, I'll introduce you to each of the three teams and I'll provide you with some useful tools, tips, and advice for applying the framework in your scaling business.

The 80/20 Rule

You've likely heard of the 80/20 rule. If you're unfamiliar with it, allow me to summarize. The 80/20 rule, also known as the Pareto Principle, simply states that 80% of the results are derived from 20% of the causes. This natural phenomenon shows up in all sorts of places. It has been observed in criminal justice with 80% of the crimes being committed by 20% of criminals. On roadways throughout the world, studies have shown that 20% of drivers are the primary cause of 80% of the road accidents. And, just as fitting for our discussion, 20% of employees are responsible for 80% of the productivity results within a business.

That statistic is worth restating: 80% of the value that's developed in your business will originate from 20% of the people. I've seen this principal reign true in business time and time again. As such, I've developed the three teams framework as a simple model that can help you leverage the Pareto Principle to your advantage. It will guide you in creating and sustaining a high-performing team in your scaling business.

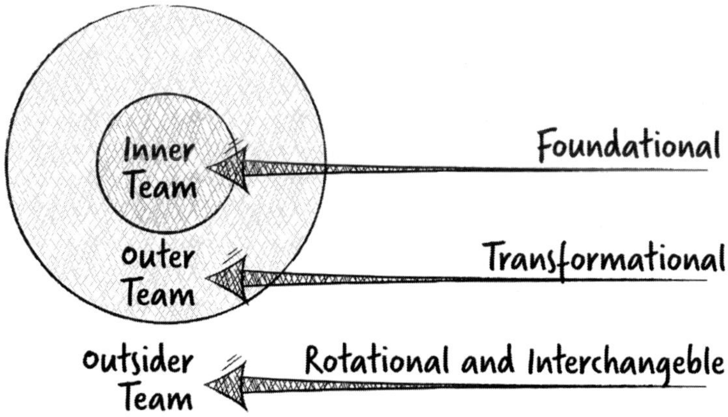

Figure 9.1 The three teams framework, composed of an inner team, an outer team, and an outsider team

The Inner Team

Your inner team is the tightest-knit group. It is the nucleus that the rest of your organization revolves around. Don't mistake the inner team simply for your leadership team or your direct reports. The inner team is not composed that way. It is very different than a cluster of roles on your organizational chart. In fact, the inner team emerges because of the beliefs and actions of the founder and the founding team members. Everything affects it—your thought patterns, upbringing, cultural background, heritage, bias, prejudice, etc. Whatever attributes you can think of is what forms and bonds the inner team.

The inner team provides a unique function to the organization. It helps to set direction and contributes to the company's predominate culture. This collective-mind paradigm dominates your business's operating rhythms.

I have a friend who developed a software product on top of Microsoft SharePoint. For many years, this company operated as a consultancy, but it finally broke into the product space. It was predominantly an Australian company, but it began to buy businesses around the globe as it executed a growth-through-acquisition strategy. Furthermore, my friend decided to move his headquarters to New York and began staffing their business with Americans.

While you wouldn't think there would be the biggest cultural divide between America and Australia, you'd be incorrect. The Australian culture tends to be much more carefree. As the Australians love to say, "No worries." This culture showed up in the new U.S. headquarters as one where the Australians enjoyed the best projects and the most high-profile assignments. It created a division between the American workers, who felt as if their jobs were to do all the work, while the Australian crew had fun and partied together. For better or worse, this inner team's operating system set the cultural norms by which the organization operated.

Your company too has an inner team that sets its culture. We often call the members of these teams "foundational" because they form the bedrock of the business. They tend to be people who are highly connected to one another because of some shared belief system, a similar worldview, long-standing business partnerships or friendships, or perhaps even just a shared sense of duty around the mission.

Another important aspect of foundational people is that they tend to stay on with your company for seven to ten years, or even longer. They are tightly connected to the mission and will serve to evangelize your product and your purpose internally and externally. They also tend to care much more about the organization and will often act in ways to protect and guide its growth. Where I began my career, it wasn't uncommon to see people who had racked up 20 years of tenure at the company. Yes; even in 2023! It was extremely common to hear people discuss product solutions by asking: "Was this in the best interests of the investor?" A devotion to your cause (and not just lip service) is a key indicator that you've developed a foundational team member.

I previously served as an expat for that same company. They sent me to their Australian subsidiary to lead a major transformation project. One of my key objectives was to transplant the culture because I was seen as a foundational member of the group. As such, my responsibility was to represent our cultural brand and to infect others with that same culture. As it turns out, the Australian team resisted my attempts and I found myself on the outer team in that office.

The Outer Team

The outer team doesn't necessarily function as a cohesive team either, but here we find what is generally the largest group of people within your company. The outer team sits just outside the inner team and functions as the workhorse of the organization. A great deal of the work happens among these people, but they are perhaps not the greatest contributors to your company's thought leadership. By thought leadership, I mean to say that the voices within these groups may be important for technical or specialist leadership, but they're unlikely to have the loudest or most impactful voices when it pertains to the soul of your company. In essence, as leaders, we are looking for the outer team to be converted to the inner team's paradigm and way of working.

The outer team is also mostly comprised of transformational people. Transformational people are those who have joined your organization because they are hoping to take their career from one place to another. I once worked as a business analyst and was looking to become a project manager. I took a low-level project management job at an app development agency because it would provide me with the opportunity to transform my career potential. I actually took a pay cut for that job because I believed it would give me the essential experience I needed to elevate my career to a much better place. The strategy worked. I was the typical transformational employee. You'll quickly recognize these types of people in your organization as well.

Transformational employees tend to stay with your company for 12 to 18 months—long enough for the transformation to occur. Once the transformation occurs, those employees will be looking for another transformation or the opportunity to dive deeper into their specialty. You should prepare for this eventuality and ensure you have a new transformational role ready for them in your company. As this person is highly skilled and very talented, you'll be motivated to continue creating opportunities for them. Additionally, you'll be strategically positioning them to become a foundational team member who propagates the culture and your vision for the world. If not, you'll be planning for their eventual exit.

The Outsider Team

Finally, we have the outsider team. As the name suggests, they are very much on the outside and will likely receive very little attention from you. This team tends to be filled with rotational and interchangeable roles.

Rotational roles are generally entry level or junior. These individuals are exploring what they want to do and have not yet developed a specialty. They are interested in learning and growing at your company. Some of them may be highly motivated and could quickly find themselves becoming transformational team members or even, eventually, foundational team members. Rotational employees initially tend to take on quite low-level, repetitive, and must-do work to increase the quality of your product and your company's ability to service its customers in a consistent manner. Often these are customer service or customer success roles.

Interchangeable roles are like rotational roles in the sense that they are defined by either repetitive or routine work. However, unlike rotational roles, this work is often specialist and is of high value to your business. It must be done correctly and with precision. Examples are outsourced software development. Companies may need to develop proprietary software that isn't mission critical, for instance, but that nonetheless requires a high degree of precision and quality.

Another great example are businesses that need to create highly detailed business processes and business requirements. Often, these jobs cannot be avoided and require interchangeable expertise to come onboard and assist the inner team with smashing down the workload. These people act as a contingent workforce, providing your business with the capability to flex up and down as necessary.

At any given time, your business may have five or so critical objectives. But things change. Some of your less critical objectives may become more important and others may need to take a back seat, meaning you'll need to change the size and makeup of your teams. This is a great opportunity to use a third-party service that can quickly provide you will skilled expertise. These will be your insourced consulting partners or perhaps your contingent workforce of contractors. As mentioned earlier, look for partners that will allow you to recruit some of their people permanently because, when you're flexing down, you don't want all your IP going out the door.

Moving High Performers to the Inner Team

To build and sustain a high-performing team, your goal should be to move the highest performing people from the lower rungs of the outer team up to the foundational team. It's not enough just to have highly skilled and highly talented people in your organization. They need to become a functional part of your cultural operating system; it should be expressed in how they show up in work every day. This is important because so much of knowledge work is a function of our interpersonal relationships; of how we connect and relate to one another. So much information is transmitted through non-verbal communication. And so much of our communication is the things left unsaid rather than the things that are said.

Your foundational team members are more likely to better communicate with each other because they have a stronger sense of shared values and beliefs and a culture that bonds them. But you must have a plan for refreshing your talent because your foundational people will not stay forever. And you already know your transformational people are only with you so long as you can keep supporting them through meaningful transformations.

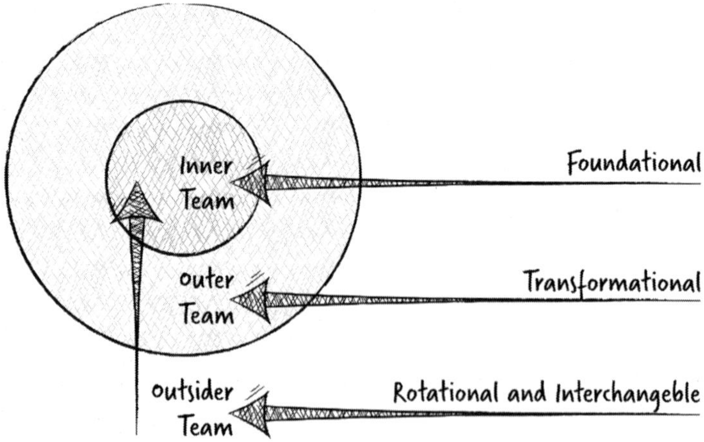

Figure 9.2 Move the outsider team towards the center.

Staffing Each Team through Talent Cycles

To build and sustain a high-performing team, you'll need to understand each person's talent cycle and how they map to each of your three teams. Every person in your business will follow the natural cycle of growth, promotion, stagnation, and decline.

Growth

Growth, in any role, no matter the specialty, is a requirement for building and sustaining a high-performing team. Your business success depends on your people growing in each role in the company. Growth in this context means that they reach the heights of their potential for a given job. You'll notice that not everyone grows at the same rate. Regardless, you should expect that your team is growing deeper in their specialty or greater in their influence.

The growth stage is characterized by a person who continues to reach new levels in your company. Their skills are being put to good use. They are flourishing as people, and more importantly, you notice they have a unique skill to create growth around them. Whether it's growth in your organization's capability or growth within a technical expertise, these individuals drive great results, they often don't require a lot of overhead and management, and they are usually headed toward promotion.

Promotion

While not every person in your company will move up to take on additional responsibility, many will experience lateral moves or more technical specialization. You should expect these transformations every 12 to 18 months.

The promotion stage in our model is where the person is given new tasks, responsibilities, and assignments that stretch them further than before. They take on increasingly complex tasks that have greater impact on the organization. Often, those who go through several cycles of growth and promotion become your foundational people. It's uncommon for people to be great leaders in your organization if they don't share some of the inner team's beliefs and ways of working, which naturally leads them to become foundational members.

Stagnation

Stagnation is the name of the game for many mature businesses. Often, these businesses are filled with people who have stopped growing and are on autopilot. For a growing or scaling business, this is unacceptable and should be avoided.

Stagnation is not necessarily the fault of the individual. It tends to happen when there aren't enough roles in your business for people to get promoted and accept increasing responsibility. However, sometimes it is the fault of the individual in not managing their career properly. These people can be a risk to your business because they require more management and more attention to keep them productive. They generally map onto the transformation and rotational roles within the outer and outsider teams. While most high performers who reach stagnation will quickly look for something else, either in your company or elsewhere, the low performers will tend to stay in stagnation simply to collect a paycheck and fly under the radar. This can be an enormous threat to your company.

Decline

Decline is when an employee has stayed in your business beyond their usefulness. This tends to happen when your business changes direction but you've not been fully cognizant of the people who should no longer be onboard.

While stagnation is the beginning of decline, decline is characterized by people who have disengaged from the organization and need to exit or be exited. While it's rare for foundational team members to find themselves here, it can happen, but it's more likely that they would have self-selected out or you would've had to fire them.

You can use the performance versus potential matrix tool outlined in Figure 9.3 below to determine what stage a person is in and review their capacity to move into a different stage.

The performance versus potential matrix is a simple three-by-three framework that helps you determine a team member's value to your organization based on their skill level and their engagement. Your goal is to identify members of your team who stand out as future leaders, growth team members, and high-impact performers. For the rest, you'll need to start thinking about their expiration date and how to eventually move them on. The truth is that your scaling product company will experience significant change, and they'll be unable to keep up.

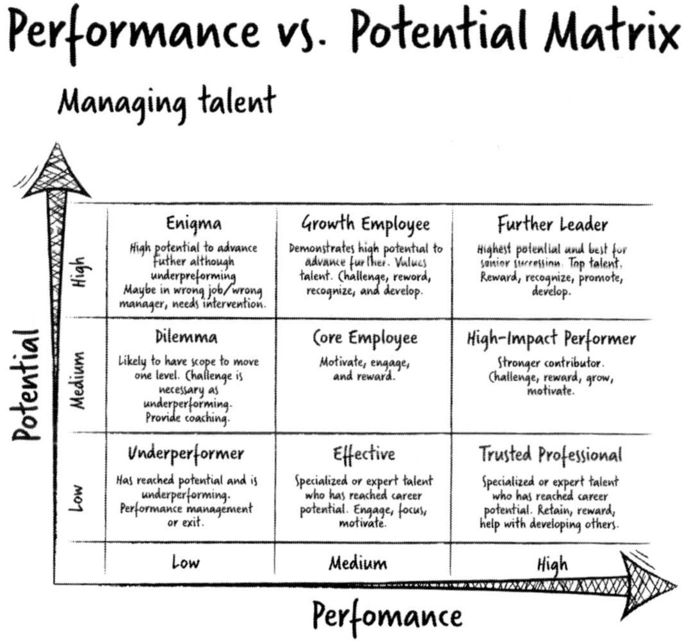

Figure 9.3 Performance vs. Potential Matrix, a tool for managing talent.

Every Role Comes with an Expiration Date

As you can see, talent isn't stagnant. Your people are constantly changing, as are your customers, the market, and technology. As a result, your business will constantly change too. Knowing that change is inevitable and constant, we must design our companies around this fact. We must acknowledge that building and maintaining a high-performance team isn't accomplished by completing a set of actions. It's much more complex and requires significant, ongoing maintenance and an awareness that many of your greatest people today will soon become former employees.

You must build in expiration dates as you observe and manage each team member throughout their life cycle. As you have come to understand, some of these expiration dates will reset as your people transition to new roles within your organization. For others, however, you will need to develop a clear plan for how to transition them out of your company.

When we first recruited Rob, he was an incredible fit for our organization. He provided a sense of fun, friendship, and connection that was lacking on our foundational team. He brought a feeling of lightness to our work and really wanted to enjoy himself. Among a group of really serious people, he changed the dynamic. It was only a couple of years into our journey together that one of our clients made a comment about Rob that helped us to realize that he was no longer a fit. My first thought was to be defensive and to want to protect him, but I thought deeply about what she said. She was right: Our collective growth, as a foundational team, was outpacing Rob. His role was starting to hold us back. We made the difficult choice to let him go.

When I started my first business, I wasn't aware of these realities. I've always been high on care toward people. What I learned was that, at the end of the day, this is an important part of building and growing a business. We have to be comfortable acknowledging that, over time, not everyone fits, and that we need to let some people go. It's obviously easier to let people go when you haven't become too close. It's always harder to let go of someone you really care about or have become friends with. This doesn't mean that you should be cold and distant from your employees. But you should prepare for the fact that they have an expiration date, and you should have already thought about their planned exit.

Leveraging Contract Roles for Planned Exits

Contract roles are a simple and effective means to build these talent off-ramps in your business. You simply utilize your network of partners to bring on great talent with clearly defined roles and end dates. This way of staffing your team has a built-in refresh button. Even those contract roles can become stagnant. Refreshing your talent in this way helps to provide clarity on whether you have the right priorities.

Contract periods allow you to build contingency into the business by leveraging your external service providers' talent pools to work on your initiatives. Rather than trying to staff up your business with many more full-time employees, a time-consuming and costly exercise, you should seek to develop a composite team of specialists.

Over time, companies develop a level of "bloat" that needs to be eradicated. Doing so gives you the chance to determine if you ever really needed to provide those capabilities at the levels you'd developed them to. Of course, if you cut too much, you risk overworking and upsetting the remaining team. Nonetheless, this practice can help you reduce your costs and streamline the business on a regular basis.

For your permanent hires, you'll need to develop a more expansive strategy for planning their exits. This could be a combination of mission-oriented discussions developed in their performance plan or targeted KPIs to help you understand how well they're accomplishing their goals. There are many "people and culture" firms that can help you develop a strategy for managing your full-time talent.

Would You Rehire?

As you can no doubt see, making sure that you have the best team comes with a lot of thinking and planning. As the CEO, you should only have three critical jobs: managing the numbers, developing the culture, and building an organization that has the right people. To make sure you've got the best people, you'll need to regularly ask yourself if you'd rehire the people you have. And don't just think about your employees; consider your partners and suppliers. Can you say, after having worked with them for six months, that you would rehire them?

With every paycheck or invoice you sign, you are rehiring those people. Don't continue to pay for less-than-ideal services or employees. If you cannot emphatically say you'd hire that person again, then I'd ask you, what is your plan to terminate them? What actions are you taking today to end your relationship with that employee or that vendor? Ensuring that you build and maintain a high-performance team requires you to think carefully about that question. One of the key secrets to getting and keeping good people is letting go of the bad ones. Winners want to work with winners. And that's as true for your partners and suppliers as it is for your employees.

Some quick takeaways from this chapter:

- Every business has three core teams: the inner, the outer, and the outsider team. Focus most of your energy on the inner team, who produce 80% of the results.
- As you mature, create cultural systems and processes that move the highest performers toward the inner team.
- Recognize that every position in your scaling business has an expiration date, and systematically plan for new assignments and exits.
- Leverage contract roles for planned role obsolescence. Don't build capability that you won't need long-term.

As your business grows and matures, don't be caught off guard when the amazing team you built early on no longer resembles their promising beginnings. Building a business is a balancing act. You need to balance your business with who you are and where you want to go while also aligning your people and ensuring they can accomplish your goals. In the next chapter, we'll look more closely at alignment and how your biggest competitor might hold the answer.

10

Keep Everyone Aligned

How alignment becomes more challenging as you grow and what to do about factions.

People require an enemy. Someone to hate. That could be a competitor or even a group within your business. Don't try to optimize for harmony; instead, optimize for productivity.

During the 1985 Geneva Summit, President Ronald Reagan and Soviet Premier Mikhail Gorbachev took a break from negotiations to take a walk. The details of what they talked about were kept secret from both the Russian and American public for years; only their private interpreters were present. In a 2009 interview with Charlie Rose and Reagan's Secretary of State George Shultz, Gorbachev disclosed that Reagan had asked him if they could set aside their differences in the event of an alien invasion.

> "President Reagan suddenly said to me, 'What would you do if the United States were suddenly attacked by someone from outer space? Would you help us?'"
>
> "I said, 'No doubt about it.'"
>
> "He said, 'Me too.'"
>
> "So, that's interesting," Gorbachev said to much laughter.

Dr. Felicitas Flade and her research colleagues published a study in the *Journal of Experimental Social Psychology* that showed how uniting against a common enemy changes the way people perceive similarity across racial and gender divides. When we come together to face a foe, it transforms how we see other people and how we categorize them.

In their experiments, they showed participants several speakers' faces or names along with the statements they made. The discussions were either about daily life or about common threats like terrorism or disease. There were two clear categories of speakers. In some studies, participants were either Black or white; in some, they were male or female, and in one study, they were Jewish or Arab. Next, participants were asked to recall who made each statement.

The researchers focused on the mistakes people made in identifying who made each statement. If, for example, one of the Arab speakers said something and the participants couldn't remember exactly who said it, would they be more likely to guess that it was one of the other Arab speakers or a Jewish speaker? That would suggest that they were categorizing speakers on the basis of being Arab versus Jewish if they thought it was another Arab speaker.

However, when the speakers were discussing a common enemy, participants made fewer mistakes based on skin color and gender. Participants were less likely to see people as Black or white or as women or men. Even when Israeli Jews and Arabs were discussing the Israeli-Palestinian conflict, they still categorized each other less because they had a common enemy. So, what does this tell us? When people are threatened by a common enemy, they categorize others into groups less often. They unite to face the threat.

Alignment Challenges Compound with Complexity

As your team grows, you will no doubt face a host of new challenges. The days of the nicely cohesive startup team are over. Your need to introduce new processes. New layers of leadership and management will no doubt sap some of the fun and enjoyment out of your growth journey. You'll have to start acting and appearing like a more well-put-together business. With this move toward maturity also comes the inevitable growth of silos. And, yes, I know its 2023 and that most digital native organizations are flat in their structure. But hierarchies still exist, even in flat structures. Hierarchies occur in nature. One person or entity is and will always be subordinate to the other. This reality creates complexity and contributes to communication overhead. Communication overhead is the proportion of time your team spends communicating with each other instead of getting productive work done. This grows in line with the amount of people in the organization (as illustrated in Figure 10.1 below). Teams experiencing communication overhead often suffer from these symptoms:

The invisible decision – No one knows how or where decisions are made, and there is no transparency in the decision-making process.

Unfinished business – Too many tasks are started but very few are carried through to the end.

Coordination paralysis – Nothing can be done without checking with a host of interconnected units.

Nothing new – There are no radical ideas, inventions, or lateral thinking. A general lack of initiative.

Pseudo-problems – Minor issues become magnified out of proportion.

Embattled center – The center battles for consistency and control against local/regional units.

Negative deadlines – The deadlines for work become more important than the quality of the work being done.

Input domination – Individuals react to inputs (i.e., whatever gets put in their in-tray) as opposed to using their own initiative.

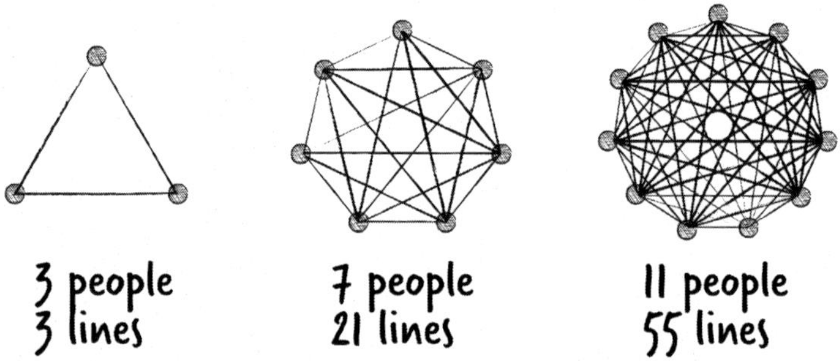

Figure 10.1 Lines of communication are exponential as you grow

As your organization scales, so does the amount of time you spend communicating with them. Eventually, they will spend almost all their time communicating with each other and very little time working independently. Each additional team member beyond a certain threshold diminishes the capacity of the group to do anything other than communicate. There is, however, a very simple solution to this problem. Reduce the size of your teams to reduce the need for communication.

Leadership Dysfunction Flows Downward

As growth and increased complexity creates communication problems, it can also create division and turf wars between the clashing personalities of your most valued leaders. Once I served as the COO for a scaling digital product business that had some ridiculous silos that emerged because of difficult interpersonal relationships. The CTO of that company lacked some basic emotional intelligence skills. He frequently steamrolled his development team's solutions, undid their work, and made them all feel stupid when he did decide to communicate with them. He had a special way of letting everyone know that he believed he could do their job better than them.

To protect his people from the brunt of his regular onslaughts, the head of design proceeded to organize his team like a design agency within the startup. I recall, on several occasions, being instructed to wait at the door of their "design studio" to have a conversation with one of the designers. But, before being able to speak with the designer, I had to run my request by the head of design.

Obviously, this setup was extremely wasteful, inefficient, and culturally destructive. On top of that, it created an unproductive, dysfunctional relationship between the development team and the design team. Designs were handed over to the development team without any real consideration of whether they could even be implemented. The developers were afraid to push back with the designers. Ultimately, we struggled to ship a product.

Team dysfunction always flows downward. If two groups in your business find it difficult to get along and work together, you only need to look at the leaders to discover the problem. Ultimately, it's your responsibility to resolve silo issues, but this will only be one of many issues that you're dealing with on a regular basis. These fundamental disagreements distract you from your mission. But unresolved differences eventually lead to compromise and inefficiency.

The places where I've witnessed this dysfunction the most are in mature companies that are exploiting a cash cow with very little regard for anything standing in their way. The name of the game here is to wring it dry while the getting is good. I've also witnessed this type of behavior on display in declining organizations that are on the brink of failure. They are at the point where they are throwing anything and everything at the wall to see what sticks. It's easy to witness these behaviors in these organizations because they are trying to stave off collapse.

The other place is in new, scaling businesses, where the founder or the CEO is focused on growth and the mission at all costs. This is fairly common when you're running toward aggressive deadlines and funding milestones. The trick here is to ensure you balance employee happiness and maintain alignment in the right places.

The Tribal Leadership Framework

There is a model we can use to help us better conceptualize these different stages of alignment and cultural evolution (or de-evolution). In the acclaimed book *Tribal Leadership*, the authors, Dave Logan, John King, and Halee Fischer-Wright, outline several stages in which your company, or parts of your company, may be. They define a tribe as 20 to 150 people that are responsible for getting work done within an organization. A tribal leader is one who builds the tribe and helps it to develop along the stages of tribe maturity. It is important to note that the best tribal leaders are responsible for leaps in productivity, profitability, and employee attraction and retention. Let's look at those stages of tribal maturity and how you might progress toward a higher stage of alignment with your team.

Stage One

In many cases, stage one isn't worth talking about. The authors describe the mentality at this stage as "life sucks." They recount stories of consulting to stage one companies that no longer exist because of accounting scandals and people coming to work with shotguns. I hope it's a safe bet to say your company will never be here.

Stage Two

Twenty-five percent of companies exist at the stage two paradigm of "my life sucks." They are passively antagonistic, laughing sarcastically and possessing a judgmental demeanor without the motivation to get involved. Large, corporate organizations often have parts of their organizations that exist at this stage. These departments tend to be procurement, accounting, and human resources. In my own consulting experience, many government agencies exist at this stage.

Stage Three

Most workplaces (49%) exist at stage three, which carries the sentiment, "I'm great, and you're not." This behavior can be observed among doctors, lawyers, and salespeople. For people at stage three, winning is personal, and they must reign supreme. They will outwork and outthink their competitors on an individual basis. The resulting mood is a collection of "lone warriors" who want help and support but who are continually disappointed that others don't have their ambition or skill. Their complaint is often that they don't have enough time or competent support because they must do the tough work.

Stage Four

This stage represents 22% of companies with a collective paradigm of "we're great." The space between "I'm great" and "we're great" is a massive chasm. People in these organizations are happy, inspired, and genuinely themselves, without the need for masks. Groups at the "we're great" stage usually have a common purpose, but they *always* have a common enemy.

Stage Five

At stage five, you'll be thinking of tree huggers and cults. These companies have created fanatical ambassadors inside and outside of their organizations. The collective mindset in these companies is, "Life is great." The people who work at these companies are not thinking about beating their competitors; rather, they are focused on changing the world. A good example of a team at stage five would be the initial team that created the iPhone. Such revolutionary technology truly did change the world.

For you, our scaling CEO, your journey is likely to be characterized by people issues across stages two and three. Increased complexity exacerbates communication issues, fueling misalignment. Your business is going to experience the pains associated with that growth. To build an outstanding team, you must be prepared to lead through the peaks and valleys of each stage. Misalignments left too long among key parts of your company will breed factions.

Manage Stage Two Factions

In my story above, I talked about two leaders who created division and two factions that could not work together. In this situation, I focused on upgrading the organization from stage two to stage three by infiltrating both factions. With the head of design, I noticed what his team liked to do: go for beers and trash the rival faction. I worked my way into those sessions, presenting myself as unthreateningly as possible, and participated in the banter without breaking my own sense of integrity. I found ways to use language to highlight their strengths and gently encouraged them to show more of their positive behavior toward our developers without reprimanding them for their bad behaviors.

It worked. My connection with the head of design allowed us to cover a lot of ground. Ultimately, our designers and our developers began to work much more closely. As the leader of your scaling company, you'll have to discover clever ways to make the factions work to serve your purpose. And that means coaching, relating, and connecting with your key leaders. And, when all else fails, know when it's time to change the people.

Leverage Stage Three Hostilities and Factions

I once worked for a brilliant manager who built one of the most diverse and cross-functional teams I'd ever seen. This leader was great at getting the most out of his people. He understood personalities and how to organize individuals to get stuff done. He appointed Barkha, the technical lead, to a role where she had direct responsibility over the technical team. Her methods were direct, sharp, and sometimes insulting. But she got results, and John knew that. Barkha was a typical stage three actor; solely out for herself. She acted in a way that boosted her own credibility and chastised anyone who could not keep up with her.

The brilliant thing about her behavior was, although people didn't like her, her forceful management generated a lot of productivity. John leveraged her to get an incredible amount of work done. You'll want to be careful with your own high-performing stage three leaders. If you unleash people like this in your organization, you must be mindful of when lines are crossed and that you never get into bullying territory.

Find a Common Enemy to Align Against

At stage two, you are most likely the enemy within the organization. People are directing their hate toward you and your leadership. At stage three, the hostilities tend to be felt between groups and within groups. Ultimately, you want most of your people focused on a common enemy that exists outside the organization—your competitors. At this stage, they are looking past your shortcomings and failures. They have stopped focusing on the animosity that exists between them. They are now linked, arm-in-arm, thinking about how they can do better than their loathed market rivals.

Promote or Hire Those Ready for Stage Four

Don't just promote the most talented or the most skillful individuals. Look to promote those who are powerful. And by powerful, I mean those people who are high on influence. The ones who, when they speak, others listen to. The ones who create change. These people are often ready for more challenging projects to improve their skills. They will help you grow and improve alignment because they are naturally wired to play a stage four game.

And remember, your stage four people don't necessarily have to be full-time employees. They could be contractors or consultants. They could be individuals seconded from universities or other businesses. The key point is that they share your passion for beating your competitors. They must be high-performing individuals who share your values and aspirations for greatness.

Some quick takeaways from this chapter:

- Growth exacerbates complexity. Expect to experience significant communication disruption as you scale.
- Plan for bringing in additional layers of management and creating new communication processes.
- The dysfunction that you tolerate among your leadership team will flow to the lowest levels of the company and undermine your business's effectiveness.
- For your scaling business, alignment can be maintained by uniting your people against the enemy outside your four walls.

Over the previous chapters, we've focused a lot on the people that you're going on this scaling journey with. But you're the one that must find and hire the right people, navigate successfully through the ups and downs of scaling, and keep your teams aligned as complexity grows. The final section will examine how you're going to get there. This chapter, a critical one, is all about you.

› REFERENCES

Flade, Felicitas, Yechiel Klar, and Roland Imhoff. "Unite Against: A Common Threat Invokes Spontaneous Decategorization Between Social Categories." *Journal of Experimental Social Psychology* 85 (2019): 103890. https://doi.org/10.1016/j.jesp.2019.103890.

Flade, Felicitas. "My Enemy's Enemy Is My Friend: Common Enemies Blur Lines Between Groups in Conflict." Society for Personality and Social Psychology. January 27, 2020. https://spsp.org/news-center/character-context-blog/my-enemys-enemy-my-friend-common-enemies-blur-lines-between.

Kaufman, Josh. "What Is 'Communication Overhead'?" The Personal MBA. Accessed December 16, 2022. https://personalmba.com/communication-overhead/.

"Reagan and Gorbachev Agreed to Fight UFOs." The Christian Science Monitor. April 24, 2009. https://www.csmonitor.com/USA/Politics/The-Vote/2009/0424/reagan-and-gorbachev-agreed-to-fight-ufos.

"Reagan and Gorbachev Agreed to Pause the Cold War in Case of an Alien Invasion." *Smithsonian Magazine.* November 25, 2015. https://www.smithsonianmag.com/smart-news/reagan-and-gorbachev-agreed-pause-cold-war-case-alien-invasion-180957402/.

PART IV

UPGRADE YOUR CAPTAIN

11

Lead Yourself

Building the right team always begins with leading yourself.

Your personal shortcomings, character flaws, and fears will shape the culture of the business.

There he was, boarding the 6 am flight to Los Angeles. He always wore a pair of black men's slip-on Skechers with a cliché, computer nerd T-shirt. It probably had some silly coding quip that most likely went over my head. He would've thought it was hilarious. I always saw him as the quintessential startup CEO. Extremely passionate about technology. Ridiculously smart. Somewhat incapable of connecting with the average human being. But he still managed to build a really solid team of engineers who, for better or worse, believed in him.

It's a long flight from Australia to Los Angeles. But then it's just a short hop over to San Francisco. He was meeting with some of the biggest names in venture capital. I can only imagine the feeling of being in the office of Andreessen Horowitz. Meeting with executives from Sequoia Capital. That ever-present sense of excitement that comes with being in the company of people who could change your life in an instant. To even have landed those meetings was a miracle. What an amazing opportunity.

He reported back to the team about how well these meetings were going. He told us about how much the investors loved our product. How impressed they were with the pitch. How many other amazing meetings were currently being scheduled with other big-name players in venture capital.

After that whirlwind tour through the Bay Area, he was off to New York and London to meet with more investors. Meanwhile, back in Australia, having lost confidence in his leadership, his team slowly began to resign. One by one, people started to head toward the door. What would cause people to walk out on their shares of what was once a promising scaleup company?

The Dysfunction Within

I've coached many business leaders who are suffering in silence. They feel trapped. They feel alone. There's so much resting on their shoulders. There's the weight of decision-making. The responsibility of spending other people's money. The pressure of running the organization. Not to mention the mental toll of their ambition. The lack of sleep and the emotional ups and downs. The second-guessing of choices they've made. Eventually, you must stop and ask yourself: Why am I doing this to myself? What's underneath it all?

I've coached many leaders that have so much unaddressed inner turmoil. The result is a trail of destruction in their personal lives and businesses. I've witnessed that it is nearly impossible to build a sustainable, high-performing, and productive team when we function from a place of internal chaos.

There was Harriet, the CIO of a global charity, who found it impossible to say no to her colleagues' requests. She suffered from an incredible inability to make decisions and was bogged down with people-pleasing behavior. Her organization worked on numerous initiatives simultaneously and rarely completed any. Then there was Victor, the high-profile tech banking executive who frequently used intimidation and fear to get his way. His tactics caused a 300-person program of work to unravel within a matter of days. Monica, a startup CEO, was so obsessed with her image and personal brand that she ignored the fact that her team hadn't shipped a single product in five years. Sean, a digital analytics executive, failed to lead his organization effectively because he was too concerned with what his subordinates thought of him. The teams he headed rarely, if ever, shipped any major programs.

Ted, the leader of a tech startup, found it incredibly difficult to take a break. His tendency was to take care of everyone on his team. To his credit, his team loved him and relied on him immensely. But eventually, he'd crack under the pressure, go on a multiday bender, and fall into a depression because of what he'd done. Everyone at work could see he was burnt out. His behavior and demeanor was grumpy, short, and callous, and he wasn't much fun to be around. What was the underlying driver of his behavior? It was the need to please his parents. You see, Ted never felt like his parents loved him. He was stuck. He kept playing out a script in his mind where he needed to rescue the people around him. Rather than taking care of himself, he spent all his energy taking care of others. And then there's you.

What impact will your dysfunctions have on your business and your success?

Much of the dysfunction we create is the result of misunderstanding or ignoring our shadow side that is lurking below the surface of our personalities. Most of the advice in this book will still be useful if you choose to ignore what's in this chapter. But if you truly want to build the right team, you will commit to working on yourself. Make it a daily habit to become the best version of you. Work on becoming the best leader that you can be.

The Shadow Self

Over time, we each learn from society that some of our behaviors, emotions, sexual desires, lifestyle choices, and relationships might be viewed as wrong or undesirable. Because these qualities disrupt social norms, others see them as "inappropriate." If someone is too "different," they are cast aside.

We naturally crave connection, acceptance, and belonging. None of us want to be viewed as the outcast of our groups, so we do whatever it takes to fit in. As children, we search for the line between socially "acceptable" and "unacceptable" and spend the greater portion of our lives trying to conform to the former.

We experience the negative consequences of society's disapproval when we break its rules. The unpleasant emotions that come with people judging, condemning, and gossiping about us can quickly become overwhelming. Although we may not realize it, we suffer for our deviances when other people observe them. We internalize society's backlash so deeply that we eventually inflict it on ourselves.

Masking the pain is the only way to remedy our suffering, and that's when our ego shows up. We create narratives about ourselves. We create a picture of who we are not and places we would never go as a person to avoid the repercussions of being an outsider. We succumb to confirmation bias, which is our tendency to interpret and ignore information in ways that confirm what we already believe. Thus, we develop a firm belief about something and unconsciously discard any information that contradicts that belief.

All of us have qualities that society deems objectionable. Some of us are prone to vanity. Others are short-tempered and naturally selfish. The model citizen is a person who manages to live up to impossible standards.

The thing that nobody wants to admit to other people is that we're all secretly not meeting those standards. Jung called the social masks we all wear our "personas." Women wear makeup, men use masculine-scented deodorant, influencers use Instagram filters, and people filter their personalities with curated photos and status updates on social media—all to mask perceived flaws and project an image of "perfection."

As Jung writes, "Unfortunately there can be no doubt that man is, on the whole, less good than he imagines himself or wants to be. Everyone carries a shadow, and the less it is embodied in the individual's conscious life, the blacker and denser it is. If an inferiority is conscious, one always has a chance to correct it. Furthermore, it is constantly in contact with other interests, so that it is continually subjected to modifications. But if it is repressed and isolated from consciousness, it never gets corrected."

Integrating Your Shadow Side

In his book, *The Laws of Human Nature*, Robert Greene gives some practical advice to help us integrate our shadow side. We must *see* our shadows, which is the most difficult step. We like to mask our darker side. Many of us feel it's strange to see this part of ourselves. The fact is, we're not whole people if we can't acknowledge and accept the ugly parts of our personality. We must fight our inclination to hide and embrace and explore who we really are. For many of us, our creative side comes from those frustrations buried deep within. Look at Elon Musk or Steve Jobs; many employees account for how difficult innovators like them are to work with.

When I was scaling my consulting business, I discovered that my shadow side was a persistent sense of boredom and an underlying laziness. I masked my laziness as some altruistic belief about letting my team take center stage to reach their potential. But, after much self-reflection, I've learned to embrace these traits. Also, finding Jim Schleckser's book, *Great CEOs Are Lazy,* helped reframe how I felt about my shadow side. I now own what I'm driven by. It's boredom. I start businesses and continually challenge myself because I will feel bored if I don't. I do find meaning in leading my team, coaching my clients, and helping my clients to be successful. But the reason behind it all is boredom.

Lead Yourself

Like Patrick, a startling number of tech leaders I've worked with severely undervalue the need to first lead themselves before they focus on leading and building the organization. John Maxwell famously coined the leadership principle: "the law of magnetism."

In his book, *The 21 Irrefutable Laws of Leadership*, he makes an important but not always obvious point: Your employees more than likely have a lot in common with you. Similar people tend to gravitate toward each other. This is true of employees as well as your friendship groups. At first glance, your employees may appear to be very different. Everyone is an individual and no two people are exactly alike; Maxwell acknowledges this. However, a few key areas people tend to focus on are age, attitude, background, values, energy, giftedness, and leadership ability.

So, without focusing first on leading yourself, you'll end up attracting a random assortment of people who share your best and *worst* qualities. Patrick, from our introduction, attracted people who shared an excellence for computer engineering. They were people who had the brightest minds for developing solutions and bringing them to life. A lesser-known quality about Patrick, which even he was unaware of, was his desire for fame and to be adored. This was behavior that Patrick hid well, but when it reared its head, it was quite ugly. Patrick failed to understand his shadow side.

Take some time now and list your top three character strengths. Now list where you tend to have problems; your character flaws. Think deeply about how your character strengths, when overused, could show up as weaknesses. And how might your character weaknesses present themselves if they were fully integrated into your personality?

Building the right team and being a good leader aren't some altruistic callings. It's simply about knowing where to take your organization and knowing how your personality and personal inclinations are geared toward generating positive gains. Sometimes, you'll have to be the bad guy. That's okay. When it's for the sake of your mission, you may need to leverage your shadow side. Its best to know your strengths and weaknesses, your shadow sides and bright sides, and capitalize on both spectrums of those qualities.

When it comes to building the right team, you'll want to look for people who round out your worst tendencies rather than add to them. Look behind the candidate's skillset and focus more deeply on their character. For example, if you're too nice, you might need to bring on people who are more ruthless.

Some quick takeaways from this chapter:

- Many leaders suffer from unaddressed inner turmoil that can cause destruction in our personal lives and businesses.
- You must understand the importance of working on yourself and becoming the best version of yourself to be an effective leader.
- By recognizing and embracing our shadow selves, we can become more authentic leaders and build better teams.

Upgrading your captain may be the final step in the Scaleup Culture Cycle, but it may also be one of the most important. The business you are growing will reflect your best and worst. With every wave of business growth and change, find new ways to know yourself better. By doing so, you're guaranteed to build a resilient scaleup.

› REFERENCES

Othon, Jack E. "Carl Jung and the Shadow: The Ultimate Guide to the Human Dark Side." HighExistence. August 7, 2020. https://highexistence.com/carl-jung-shadow-guide-unconscious/.

Conclusion

You've just finished *The Resilient Scaleup*. You're probably thinking about your business and the scaling journey that you're preparing for. Or perhaps you're in the middle of your journey and the advice and wisdom on these pages have opened your eyes. You probably have lots of ideas and are ready to start implementing them.

I suggest that you review your business strategy and write down all the ideas from this book that will support your scaling efforts. As you've learned in these chapters, just following someone else's playbook will not ensure success for you. You must develop your own, tailored playbook that will help you reach your goals.

Your single hardest job in scaling your company will be matching your team to your strategy and your stage of business. Remember the cultural cycle as you progress through your scaling journey. Understand your business and remember that it's going to evolve and change. You will learn more about yourself as your wants and needs grow and shift. To continually bring on the right people, you need to be in touch with yourself as these changes occur.

Secondly, you must ask yourself: Who am I going with? You must have a crystal-clear picture of what your A players look like.

Finally, you must answer the question: How am I going to get there? Reflect on yourself—both your strengths and your weaknesses—and ensure your style of leadership matches your strategy and goals.

Thank you for reading *The Resilient Scaleup*. If you enjoyed this book (or even if you didn't), please visit the site where you purchased it and write a brief review. Your feedback is important to me and will help other readers decide whether to read the book too.

References

"5 Signs That Apple Is a Cult." The Week. January 8, 2015. https://theweek.com/articles/474359/5-signs-that-apple-cult.

Bock, Laszlo. 2015. *Work Rules!: Insights From Inside Google That Will Transform How You Live and Lead.* New York; Boston, Twelve.

Feloni, Richard. "Why Google CEO Larry Page Personally Reviews Every Candidate the Company Hires." Business Insider. Accessed December 16, 2022. https://www.businessinsider.com/google-ceo-larry-page-on-hiring-2015-4.

Flade, Felicitas, Yechiel Klar, and Roland Imhoff. "Unite Against: A Common Threat Invokes Spontaneous Decategorization Between Social Categories." *Journal of Experimental Social Psychology* 85 (2019): 103890. https://doi.org/10.1016/j.jesp.2019.103890.

Flade, Felicitas. "My Enemy's Enemy Is My Friend: Common Enemies Blur Lines Between Groups in Conflict." Society for Personality and Social Psychology. January 27, 2020. https://spsp.org/news-center/character-context-blog/my-enemys-enemy-my-friend-common-enemies-blur-lines-between.

"The High, High Cost of Recruiting Developers – Dice Insights." Accessed December 16, 2022. https://insights.dice.com/employer-resource-center/high-cost-recruiting-developers/.

"How a Software Firm Achieved a $25 Million Exit by Embracing Scaling Up." ScaleUps. February 17, 2022. https://scaleups.com/how-a-software-firm-achieved-a-25-million-exit-by-embracing-scaling-up/.

"Industry Market Research, Reports, and Statistics." IBISWorld. Accessed December 16, 2022. https://www.ibisworld.com/industry-statistics/market-size/employment-recruiting-agencies-united-states/.

Isaac, Mike. "Inside Uber's Aggressive, Unrestrained Workplace Culture." *The New York Times*. February 23, 2017. https://www.nytimes.com/2017/02/22/technology/uber-workplace-culture.html.

Kaufman, Josh. "What Is 'Communication Overhead'?" The Personal MBA. Accessed December 16, 2022. https://personalmba.com/communication-overhead/.

Kotter, John P. "Leading Change: Why Transformation Efforts Fail." *Harvard Business Review*, January 2007. Accessed February 3, 2023. https://www.lifelongfaith.com/uploads/5/1/6/4/5164069/___leading_change_-_kotter.pdf

"The Law of Reversed Effort: The Harder You Try, the Harder You Fall." Big Think. May 14, 2022. https://bigthink.com/neuropsych/law-reversed-effort/#:~:text=The%20Law%20of%20Reversed%20Effort%20was%20first%20coined%20by%20the,the%20less%20we%20shall%20succeed.

Lucas, Suzanne. "What Is Topgrading Interviewing? Our Favorite Tips." Workable. August 26, 2022. https://resources.workable.com/tutorial/topgrading-interviewing#:~:text=What%20is%20topgrading%20application%3F,candidates%20for%20a%20given%20position.

Lute, Julian. "Company Culture – Meaning, Benefits, and Strategies." Great Place To Work®. Accessed December 16, 2022. https://www.greatplacetowork.com/resources/blog/company-culture-meaning-benefits-and-strategies.

Othon, Jack E. "Carl Jung and the Shadow: The Ultimate Guide to the Human Dark Side." HighExistence. August 7, 2020. https://highexistence.com/carl-jung-shadow-guide-unconscious/.

"Planning Fallacy." The Decision Lab. Accessed December 16, 2022. https://thedecisionlab.com/biases/planning-fallacy.

Quilty, David. "What Is Employee Value Proposition (EVP)?" Firstup. December 7, 2022. https://firstup.io/blog/what-is-evp/.

"Reagan and Gorbachev Agreed to Fight UFOs." The Christian Science Monitor. April 24, 2009. https://www.csmonitor.com/USA/Politics/The-Vote/2009/0424/reagan-and-gorbachev-agreed-to-fight-ufos.

"Reagan and Gorbachev Agreed to Pause the Cold War in Case of an Alien Invasion." *Smithsonian Magazine.* November 25, 2015. https://www.smithsonianmag.com/smart-news/reagan-and-gorbachev-agreed-pause-cold-war-case-alien-invasion-180957402/.

"The Real Cost of a Bad Hire." The Mason Group. September 13, 2018. https://www.masongroup.ca/employers/real-cost-bad-hire/.

"Role of the Board Topgrading Organization – Capital Solutions Bancorp." Accessed December 16, 2022. http://capitalsolutionsbancorp.com/files/pdf/Topgrading_The_Organization.pdf.

Segal, David. "Apple's Retail Army, Long on Loyalty but Short on Pay." *The New York Times.* June 23, 2012. https://www.nytimes.com/2012/06/24/business/apple-store-workers-loyal-but-short-on-pay.html?ref=business&pagewanted=all.

Sull, Donald. "Why Good Companies Go Bad." *Harvard Business Review.* August 1, 2014. https://hbr.org/1999/07/why-good-companies-go-bad.

"The Surprising Story of the Sydney Opera House's Construction – Google Arts & Culture." Google. Accessed December 16, 2022. https://artsandculture.google.com/story/the-surprising-story-of-the-sydney-opera-house-39-s-construction/CAVhiDXG4On-xA?hl=en.

Teasdale, Sheila. "Culture eats strategy for breakfast!" *Journal of Innovation in Health Informatics* Vol 10, no. 4 (2002):195-196. Accessed January 3, 2023. DOI:10.14236/jhi.v10i4.259.

Thomas, James. "Bullet Holes & Bias: The Story of Abraham Wald." Mcdreeamie's Meducational and Medhistorical Musings. July 8, 2019. https://mcdreeamiemusings.com/blog/2019/4/1/survivorship-bias-how-lessons-from-world-war-two-affect-clinical-research-today.

Thomson, Jonny. "The Law of Reversed Effort: The Harder You Try, the Harder You Fall." Big Think. May 14, 2022. https://bigthink.com/neuropsych/law-reversed-effort/#:~:text=The%20Law%20of%20Reversed%20Effort%20was%20first%20coined%20by%20the,the%20less%20we%20shall%20succeed.

"The True Costs of Employee Turnover." Built In. Accessed December 16, 2022. https://builtin.com/recruiting/cost-of-turnover.

Wharton, S. Thomas. "The 5 Reasons New Hires Fail in the First 18 Months." Miles LeHane. Accessed December 16, 2022. https://www.mileslehane.com/blog/the-5-reasons-new-hires-fail-in-the-first-18-months.

"Why Good Companies Go Bad." *Harvard Business Review*. August 1, 2014. https://hbr.org/1999/07/why-good-companies-go-bad.

"Why Is Accountability Important in the Workplace?" PowerDMS. Accessed December 16, 2022. https://www.powerdms.com/policy-learning-center/why-is-accountability-important-in-the-workplace.

About the Author

I'm a startup founder, writer, and podcast host who is determined to help scaling product businesses build the right teams. I've worked with many different startups, three of which I co-founded and two of which scaled to employ hundreds of people.

I've been obsessed with technology since my dad bought me a CD burner in 1999! When I finished my undergraduate degree, I joined a technology leadership program at the leading mutual fund company Vanguard, near Philadelphia, Pennsylvania. While in that program, I discovered I loved people, building high-performing teams, and creating innovative digital products.

I believe serendipity is life's sweetest gift, and it led me to Australia over ten years ago. It's why I spend most of my time introducing innovative product businesses to digital product experts. You never know what person or project will come along and change everything.

I've worked with Australia's Bendigo Bank, and I've built a digital product to keep Australians safe overseas. I've helped a global NGO, the World Wide Fund for Nature, refresh its digital product portfolio. My proudest achievement was helping a property tech scaleup to save millions by developing a new digital platform that helped people to love renting, owning, and managing property.

My wife and I, and our daughter, are collectively citizens of Australia, Canada, and the United States. We work, live, play, and invest in property across all three countries.

Acknowledgments

To my father-in-law, Michele Lavoie. This book could not have happened if you hadn't allowed me to take over your basement office while we were in Canada. Having that space all to myself allowed me to organize my thoughts and make so much progress on this book. I also want to thank you for setting me up in your beautiful chalet in Saint-Charles-de-Bourget. The chalet you've built is a real treasure and I'm enjoying making so many memories there. The beauty of the forest and the solitude were exactly what I needed to write this book.

To Tim Stroh and Ben Le Ralph, thank you for your research, experience, and insights. Chapter four was a missing piece of this book. Once I came across the article on our website, I knew that piece had a place in this book. Thank you both for your thoughts and hard work in creating those models. I hope this book helps extend the reach of your insights and that startups and scaleups will be more successful because of them.

To Priya Panikkar and Lu Bailey, you both took a chance on me at the start of my career at Old Dominion University and Sentara Health. I sometimes retrace my career steps and marvel at how far I've come. You both are forever a part of my story. Please know that your willingness to help an ambitious young college kid will have ripples of impact on future generations.

A special thanks to Mac Korasani for being a best friend through thick and thin. There have been so many lessons we've learned together over these years. We've pushed each other to grow personally and professionally. We've started great businesses and failed at them too! These years of learning and growth are always worth it when you have a good friend.

I want to acknowledge my amazing wife and partner in life and business. Writing this book has sometimes felt impossible. I had no idea how much time and mental energy writing a book would take. Without you, this book would have never happened. Your encouragement has propelled me when I've been unmotivated and feeling defeated. Your ability to organize, prioritize, and schedule has been immensely helpful in keeping me on track. I'm thankful to have you in my life. I'm looking forward to accomplishing so much more together with you.

Made in United States
North Haven, CT
04 November 2026

21772135R00092